Personal and Scientific Reminiscences
Tributes to Ahmed Zewail

Personal and Scientific Reminiscences
Tributes to Ahmed Zewail

Editors

Majed Chergui *(École Polytechnique Fédérale de Lausanne, Switzerland)*
Rudolph A Marcus *(California Institute of Technology, USA)*
John Meurig Thomas *(Cambridge University, UK)*
Dongping Zhong *(The Ohio State University, USA)*

NEW JERSEY · LONDON · SINGAPORE · BEIJING · SHANGHAI · HONG KONG · TAIPEI · CHENNAI · TOKYO

Published by

World Scientific Publishing Europe Ltd.

57 Shelton Street, Covent Garden, London WC2H 9HE

Head office: 5 Toh Tuck Link, Singapore 596224

USA office: 27 Warren Street, Suite 401-402, Hackensack, NJ 07601

Library of Congress Cataloging-in-Publication Data
Names: Chergui, Majed, editor.
Title: Personal and scientific reminiscences : tributes to Ahmed Zewail / edited by Majed Chergui
  (École Polytechnique Fédérale de Lausanne, Switzerland), Rudolph A. Marcus
  (California Institute of Technology, USA), John Meurig Thomas (Cambridge University, UK),
  Dongping Zhong (The Ohio State University, USA).
Description: New Jersey : World Scientific, [2017]
Identifiers: LCCN 2017026156 | ISBN 9781786344359 (hc : alk. paper) |
  ISBN 9781786344632 (pbk : alk. paper)
Subjects: LCSH: Zewail, Ahmed H. | Chemistry, Physical and theoretical. | Chemists--Egypt.
Classification: LCC QD455 .P47 2017 | DDC 540.92--dc23
LC record available at https://lccn.loc.gov/2017026156

British Library Cataloguing-in-Publication Data
A catalogue record for this book is available from the British Library.

Copyright © 2018 by World Scientific Publishing Europe Ltd.

All rights reserved. This book, or parts thereof, may not be reproduced in any form or by any means, electronic or mechanical, including photocopying, recording or any information storage and retrieval system now known or to be invented, without written permission from the Publisher.

For photocopying of material in this volume, please pay a copying fee through the Copyright Clearance Center, Inc., 222 Rosewood Drive, Danvers, MA 01923, USA. In this case permission to photocopy is not required from the publisher.

Desk Editors: Anthony Alexander/Jennifer Brough/Koe Shi Ying

Typeset by Stallion Press
Email: enquiries@stallionpress.com

Preface

Apart from being a brilliant scientist, Ahmed Zewail possessed many other noteworthy qualities.

He was a highly successful and innovative educator, an outstanding lecturer, a creator of a new science city, a skilful negotiator, a successful fund-raiser, and an inspiring mentor and research supervisor.

He was an Ambassador for Science, a post he executed with distinction, and had the stature of a Statesman.

In addition to all these qualities, he possessed the gift of friendship: all those he came within his ambit were profoundly influenced by him.

He also had a great sense of humor and an infectious laugh that galvanized his colleagues and friends to undertake worthwhile initiatives.

He was also a loving husband and a devoted family man.

He was a faithful son of, and conferred great credit upon, his native country, Egypt.

These qualities and more are elaborated in this memorial volume, which we deem to be a worthy tribute to a unique human being.

Majed Chergui
Rudolph A. Marcus
John Meurig Thomas
Dongping Zhong
Editors

List of Contributors

Fred C. Anson
California Institute of Technology
Pasadena, CA, USA

Spencer Baskin
California Institute of Technology
Pasadena, CA, USA

Wolfgang Baumeister
Max-Planck-Institute of Biochemistry
82152 Martinsried, Germany

Amyand David Buckingham
University Chemical Laboratories
University of Cambridge
Lensfield Rd.
Cambridge CB2 1EW, UK

Majed Chergui
École Polytechnique Fédérale de Lausanne
Lausanne, Switzerland

Charlie Campbell
University of Southern California
Los Angeles, CA, USA

David C. Clary
Magdalen College
Oxford OX1 4AU, UK

Marshall Cohen
California Institute of Technology
Pasadena, CA, USA

Michael J. Collins
University College
Oxford OX1 4BH, England, UK

Christopher M. Dobson
University of Cambridge
Cambridge CB2 1EW, UK

Marcos Dantus
Michigan State University
East Lansing, MI USA

Paul E. Dimotakis
California Institute of Technology
Pasadena, CA, USA

Norbert D. Dittrich
The Welch Foundation
Houston, Texas

Jack Dunitz
Laboratory of Organic Chemistry
ETH HCI H333
CH 8093, Switzerland

Peter Edwards
Inorganic Chemistry Laboratory
South Parks Rd, University of Oxford
Oxford OX1 3QR, UK

Harry B. Gray
California Institute of Technology
Pasadena, CA, USA

Dudley Herschbach
Harvard University
Cambridge, MA, USA

Archie Howie
Cavendish Laboratory
University of Cambridge
JJ Thomson Avenue
Cambridge CB3 0HE, UK

Colin Humphreys
University of Cambridge, UK

Noel S. Hush
The University of Sydney
NSW 2006 Australia

Joshua Jortner
School of Chemistry, Tel Aviv University
Ramat Aviv, 69978 Tel Aviv, Israel

Roger Kornberg
Stanford University
Stanford, CA, USA

Malcolm Longair
Cavendish Laboratory
University of Cambridge
JJ Thomson Avenue
Cambridge CB3 0HE, UK

Jörn Manz
Freie Universität Berlin, Germany
Shanxi University, Taiyuan, China

Rudolph A. Marcus
California Institute of Technology
Pasadena, CA, USA

Paul Midgley
University of Cambridge
Cambridge, CB3 0FS, UK

Chad A. Mirkin
Northwestern University
Evanston, IL, USA

Shaul Mukamel
University of California, Irvine
Irvine, CA 92697-2025, USA

Bengt Nordén
Chalmers University of Technology
Gothenburg, Sweden

Ahmed Okasha
3, Shawarby Street
Kasr EI Nil
Cairo, 11121, Egypt

David Phillips
Department of Chemistry
Imperial College London, UK

Juergen Plitzko
Max-Planck-Institute of Biochemistry
82152 Martinsried, Germany

Martin Pope
Department of Chemistry
New York University
Washington Square NY 100012, USA

Jehane Ragai
The American University in Cairo
New Cairo, Cairo
Cairo Governorate 11835, Egypt

Dmitry Shorokhov
California Institute of Technology
Pasadena, CA, USA

Villy Sundström
Lund University, Sweden

John Meurig Thomas
University of Cambridge
Cambridge CB3 0FS, UK

Dongping Zhong
Ohio State University, Columbus, USA

Dema Faham
871-Winston Ave.
San Marino, CA, 91108
USA

Contents

Preface   v
List of Contributors   vii

Chapter 1	Ahmed Zewail: Ultra-Scientist and Citizen *Dudley Herschbach*	1
Chapter 2	Ahmed Zewail: Larger than Life *Roger Kornberg*	15
Chapter 3	Stories from the Round Table *Marshall Cohen, Charlie Campbell and Rudolph A. Marcus*	17
Chapter 4	Ahmed Zewail: Great Scientist, Dear Friend *Harry B. Gray*	25
Chapter 5	In Memory: Ahmed Zewail *Jack Dunitz*	29
Chapter 6	Some Personal Recollections of Ahmed Zewail *Fred C. Anson*	33

Chapter 7	My Friend Ahmed and I Bengt Nordén	37
Chapter 8	A Tribute to Ahmed H. Zewail David C. Clary	47
Chapter 9	Ahmed Zewail — A Great Scientist and Inspiring Friend Amyand David Buckingham	55
Chapter 10	The Pyramid Builder Majed Chergui	61
Chapter 11	A Remembrance of Ahmed Zewail Chad A. Mirkin	73
Chapter 12	4D Electron Tomography: Some Recollections of the Summer of 2000 Wolfgang Baumeister and Juergen Plitzko	79
Chapter 13	Anatomy of a Friendship and Collaboration John Meurig Thomas	87
Chapter 14	Ahmed the Explorer Martin Pope	105
Chapter 15	Ahmed Zewail: Advancing Chemistry Norbert D. Dittrich	107
Chapter 16	Four Decades in the Sub-basement — Walks of Life with Ahmed Zewail Spencer Baskin	113
Chapter 17	Ahmed Zewail: An Honor to Egypt and Fellow Countrymen Jehane Ragai	125

Chapter 18	Timing with Light *Archie Howie*	133
Chapter 19	Ahmed Zewail — A Towering Visionary *Colin Humphreys*	137
Chapter 20	A Glimpse of the Evolution of Adiabaticity *Noel S. Hush*	141
Chapter 21	Ahmed Zewail: Science and Scientist *Joshua Jortner*	153
Chapter 22	Brief Encounters with Ahmed Zewail *Malcolm Longair*	161
Chapter 23	"Peter: You Have Taught Me that Electrons are Blue!" *Peter Edwards*	167
Chapter 24	How I Lost My Funding to Zewail *Shaul Mukamel*	175
Chapter 25	The Brilliance of Ahmed Zewail *Paul Midgley*	181
Chapter 26	Ahmed H. Zewail: Remembering a Hero and Friend in Science *Jörn Manz*	187
Chapter 27	Ahmed Zewail: A Reminiscence *Paul E. Dimotakis*	195
Chapter 28	"Stop All the Clocks" *Dmitry Shorokhov*	197

Chapter 29	My Time with a Giant *Dongping Zhong*	207
Chapter 30	Ahmed Zewail — An Inspired and Inspirational Scientist and Man *David Phillips*	221
Chapter 31	My Memories of Ahmed Zewail — From a Snowy Northern Sweden to the Nobel Prize *Villy Sundström*	229
Chapter 32	The Physical Basis of the Amyloid Phenomenon *Christopher M. Dobson*	237
Chapter 33	High-Intensity Mentoring and Excitement from Ahmed Zewail *Marcos Dantus*	251
Chapter 34	BCH-codes are (In Fact) Good! *Michael J. Collins*	257
Chapter 35	Ahmed Zewail: A Reminiscence *Ahmed Okasha*	267
Chapter 36	Goodbye My Love *Dema Faham*	275

Additional List of Obituaries   283

Chapter 1

Ahmed Zewail: Ultra-Scientist and Citizen

Dudley Herschbach*

To my regret, I met Ahmed Zewail in person only a few times. We first met in Spring 1976, when he arrived at Caltech, newly appointed as an assistant professor. I was visiting that semester, on sabbatical leave from Harvard. Ahmed impressed me as very bright, cheerful, and witty, brimming with ideas and zest. About a decade later, although we had not met again, I was intrigued by his innovative papers. Particularly instructive was his exposition of the role of coherence in laser spectroscopy.[1] Moreover, using molecular beams and ultrafast lasers, Ahmed had developed a pump-probe technique that mapped, in "real time," trajectories of products from photoinduced reactions,[2] exemplified by $h\nu$ + ICN $\rightarrow$ I + CN. At first, the method was limited to such unimolecular reactions, initiated by photodissociation. Bimolecular reactions seemed inaccessible because means were lacking to define a time zero.

That limitation was soon overcome, as I learned at a workshop held at Jerusalem in November 1986, devoted to stereochemical aspects of chemical dynamics.[3] Ahmed was not there, but Dick Bernstein had become a devoted apostle. A leading pioneer in molecular beam studies of chemical

---

*dherschbach@gmail.com

dynamics, Dick had, some years before, moved to the University of California at Los Angeles. Gleefully, he said he was "now a post-doc with Ahmed." In that collaboration, Dick made a key contribution by advocating a technique then recently developed by Curt Wittig at the University of Southern California.[4] It exploited the inherent mutual orientation of molecules in a weakly linked van der Waals adduct. The prototype case formed an XH ... OCO adduct in a supersonic molecular beam. Then photolysis of the hydrogen halide initiated the bimolecular reaction, H + OCO → OH + CO, propelling the H atom into the nearby carbon dioxide. The collision was roughly collinear by virtue of the precursor geometry. Zewail's team at Caltech used a picosecond laser pulse to photodissocate the HX, followed by a delayed picosecond laser probe of the OH product.[5] Thereby, they were able to clock the formation and decay of the HOCO reaction complex. The key feature was that the first pulse established the zero of time for the bimolecular reaction, enabling a series of probe pulses to record the "real-time" evolution of the correlated products.

Soon Ahmed was able to make use of femtosecond lasers, initiated by Chuck Shank and colleagues at Bell Labs. That greatly enhanced the time resolution, launching the awesome saga of femtochemistry. The name was suggested by Dick Bernstein; he and Ahmed published in 1988 a fine overview of the evolution and prospects of femtochemistry.[6] Ahmed baptized his prolific labs FEMTOLAND I, II, III and joyfully spoke of his "Femtosecond Dream" and "Femtocopia." The overview sparkled with evangelical fervor, concluding with: "This happy marriage of ultrafast lasers and chemistry promises an exciting future for this field of real-time molecular reaction dynamics."

As part of its centennial year, Caltech held a three-day symposium to celebrate Linus Pauling's 90th birthday on February 28, 1991. A year before, Ahmed had been appointed to the newly endowed Linus Pauling Professorship for Chemical Physics. On that occasion, he mentioned with a smile that two days before he had become half as old as Pauling. Aptly, the symposium extolled both Linus and Ahmed. The lectures became a book, *The Chemical Bond: Structure and Dynamics*, edited by Ahmed.[7] In his preface, he proclaimed that "as X-rays introduced the *distance* scale for molecular structure," now the advent of femtochemistry has "introduced the *time* scale for the dynamics of the chemical act itself — the transition states between reagents and products." Five chapters commemorated structure, two of them from lectures by Pauling, and the others by Alex Rich, Max Perutz, and Francis Crick. Four chapters dealt with dynamics: three from lectures by George Porter, John Polanyi, and me. The fourth chapter, not presented as a lecture, was the sparkling evangelical overview that Ahmed had published with Dick Bernstein,[6] brought up to date. Dick had died in 1990; that chapter abides among his definitive legacy to chemical dynamics.[8]

Like many physical chemists of my vintage, I am among the many scientific grandchildren of Pauling. My Ph.D. mentor was one of his most distinguished scientific sons, E. Bright Wilson. I was particularly glad to take part in the Pauling symposium because, 33 years earlier, I first visited Caltech at the behest of Linus, when he was Chairman of the Chemistry Department. Back then I gave, for the first time, a talk about my fledgling plans and hopes for molecular beam studies of chemical reaction dynamics. My lecture at the Paulingfest sampled what grew out of those plans. Particularly, I emphasized how the role of *electronegativity* — a favorite theme of Linus — nicely explains some striking aspects of elementary reactions. Ahmed protested that I had made it too simplistic. Perhaps so, yet it seemed to capture the essence, emulating Linus. I pointed to a photo, shown here as Figure 1.1, which graced the dust jacket of *The Chemical Bond* book. It surely exemplified, in a simplistic way, his favorite concept of *coherence*!

A few months later, Linus, Ahmed, and I met again, to celebrate the award of the Pauling Medal to Rudolph A. Marcus. Ahmed gave a talk on the $Br + I_2$ reaction, which proceeds via a "sticky" collision complex.[9] His new FEMTOLAND II apparatus made it possible to determine the lifetime of the complex to be ~50 ps. My talk, apropos of the 1991 presidential election campaign, was titled "How do electrons choose between candidate reaction products?" I passed out a ballot listing two candidates for each of the eight reactions. Molecular beam experiments had confirmed the predominant products, mostly "winners by a landslide." But few of the voters chose more than 3/8 of the winners, although nearly all had Ph.D.s in

Figure 1.1. Linus Pauling and Ahmed Zewail strolling on the Caltech campus with *coherent* strides.

chemistry. To my dismay, Ahmed and Rudolph each got only 2/8! Choices made on the basis of energetic or statistical considerations proved misleading. Again, Pauling-like heuristic reasoning predicted the winners, by simply taking account molecular orbital asymmetry arising from differences in electronegativity. Afterward, Ahmed sent me a brief letter, saying (diplomatically?) that his students "didn't like molecular orbitals."

In November 1992, Ahmed visited Cambridge to deliver a joint Harvard–MIT physical chemistry seminar. As well as giving a scintillating talk on femtochemistry, he romped through visits with faculty, students, and postdocs, the day at Harvard before his talk and at MIT the day after. I had the opportunity to engage in a wide-ranging discussion with him for two hours in the morning and another four hours during and after dinner. Throughout, Ahmed was delightfully vivacious, both in describing his work and asking many searching questions. He was especially curious about a whimsical verse I had hung up near my desk (see Figure 1.2). I had composed it after a seminar given at Harvard in the early 1980s by Geraldine Kenney-Wallace. At Toronto, in 1974, she

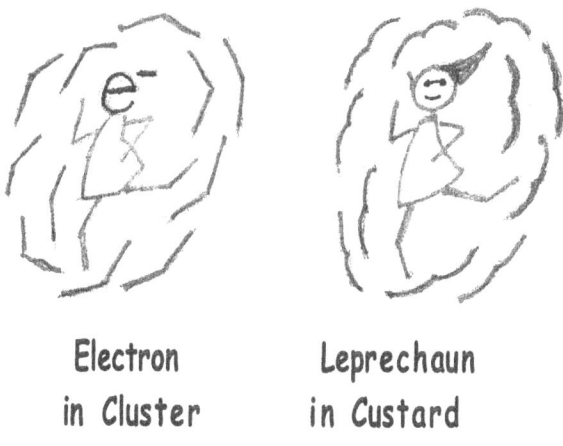

These frisky little critters are so mysterious,

They've driven many scholars quite delirious.

But now their fleeting rites and most private sighs

Are spied upon by stroboscopic laser eyes.

This is rude, yet met with grace by the merry mites,

They dance still faster to confound our throbbing lights.

Figure 1.2. Verse inspired by solvated electrons tickled by a pulsed laser.

had set up an ultrafast laser that enabled her to elucidate the formation of solvated electrons. Back then, "ultrafast" meant picosecond. The format of my verse mimics that in a booklet of poems by Robert Williams Wood (1868–1955) titled *How to Tell the Birds from the Flowers* (1917). Wood was an outstanding physicist, celebrated for his work on spectroscopy, physical optics, infrared and ultraviolet photography, ultrasonics — and for his humorous enterprises. I had composed several poems using Wood's format, which required coming up with two sketches with titles that are similar in looks and sound. Despite some effort, I had failed to concoct anything tractable for femtochemistry. Ahmed was gracious in accepting my apology. In turn, I noted that the words of the poem would apply to his femtochemistry, if the final line were changed to read: "They dance faster but cannot elude our throbbing lights."

At dinner, among things prompted by Ahmed's seminar, there came up a question about a quote he had shown on his last transparency slides (no PowerPoints yet!): "What good is a newborn baby?" It was attributed to Michael Faraday, responding to the query: "What will it be good for?" which greeted his discoveries in electromagnetism. As I was a life member of Friends of Benjamin Franklin (akin to Baker Street Regulars for Sherlock Holmes), I mentioned that the newborn baby response was made by Franklin in 1783, after observing in Paris the first flight of a hot-air balloon. The following week, Ahmed sent a letter asking for more information about the newborn baby quote. I obliged by sending copies of two scholarly articles by historians who had explored the question.[10] I mention this minor episode because of Ahmed's avid interest. Years later, he evidently had made a very thorough study of Franklin. Ahmed gave a splendid talk, "Franklin's Vision," at the Tercentenary celebration of Franklin's birth held in 2006 at the American Philosophical Society in Philadelphia.[11] In his office, Ahmed had Franklin's bust, "as a daily reminder of what it means to be a scientist in service of society and a citizen of the world at large."

Shortly after his visit, Ahmed sent a preprint of a theoretical paper, treating femtosecond dynamics on model potentials.[12] It mapped out the motion of a wave packet and its dispersion on repulsive, attractive, and barrier-crossing potentials, obtaining estimates of reaction times and relating them to parameters of the dynamics. A key issue was the time scale over which wave packets remain coherent or localized. The analysis, replete with well-designed diagrams and easy-to-follow equations, demonstrated that, on the time scale of femtoseconds to picoseconds, the wave packet spreading is relatively insignificant. This paper, a contribution to a Festschrift celebrating my 60th birthday, included a fulsome dedication, concluding: "We wish him the best and expect the Herschbachian contributions to continue with *every femtosecond full of joy*!" My reply noted that 60 years contains 3 moles of femtoseconds.

In February 1993, I visited Caltech again to give the Pauling Lecture, hosted by Ahmed. On my arrival, he graciously thanked me for the congratulatory message I had sent him on the announcement of his Wolf Prize win. Then, laughing about the 1991 ballot, he brought up the role of molecular orbitals and electronegativity. He even praised the exposition of it in my chapter in *The Chemical Bond* book.[7] Also, he had evoked a frontier molecular orbital in discussing the $Br + I_2$ reaction.[13] I had brought along the ballot. His colleague, Jack Roberts, had heard about it and asked to vote; he scored 6/8. Ahmed's secretary, who never studied a chemistry course, also voted and scored 4/8, pleased that her score equaled the sum of Ahmed and Rudolph.

As well as discussing current experiments in our labs, Ahmed was eager to hear stories about Linus, especially some told to me by Bright Wilson.[14] Ahmed, like me, had found the 1935 text by Pauling and Wilson, *Introduction to Quantum Mechanics*, unexcelled as an introduction (still in print!). Bright, as his Teaching Fellow, took notes on Pauling's lectures. Bright recalled that Linus, when responding to a question, would often pretend that he had not considered it before. He would work out the mathematical solution on the blackboard, step-by-step, with commentary implying it was a fresh excursion. Bright knew that Linus had carefully prepared the derivation before class. Bright also told me that, although he did not remember which of them had written various parts of the book, those written by Linus were turned out at a steady pace, without any revision. Several of Ahmed's co-authors on papers have testified that he too, in writing as in speaking, was remarkably adept and fluent.

Dinner the evening before my lecture was the last occasion during which I saw Linus. He told a story Ahmed enjoyed and had not heard before. Linus said that one day, just 15 minutes before his class, the phone rang. It was Ava Helen, his wife, sounding very distressed. The woods nearby were on fire, and the wind was blowing toward their house. Linus said he replied, "Sorry, darling — I can't help; I have to go teach my class." He did send a postdoctoral fellow to spray the house with a garden hose, and fortunately, the wind shifted so only Ava Helen's nerves were singed. With the look and tone of puzzled innocence, Linus concluded by saying, "But, you know, my wife wouldn't speak to me for weeks." His daughter Linda, also at the dinner, confirmed the story. Incongruous as it is, this story did emphasize the high priority Pauling gave to teaching. For him, research and teaching were kindred efforts, to be pursued with the same zest and devotion.

My AZ file has a copy of a letter I had sent to Ahmed (dated February 24, 1993), written with a quill-like calligraphy pen, to emulate Franklin; it begins: "Let me renew my thanks ... Caltech in the time of Pauling was the Mecca for structural chemistry — you are making it so for reaction dynamics in our era." Many congratulatory messages followed, most often by phone (no others by quill pen), spurred usually by papers or preprints Ahmed had sent, or sometimes by awards he had received. Most anticipated and special was his Nobel Prize, awarded in 1999. Ahmed was then 53 years old, the same age as Pauling when he received his Chemistry Nobel in 1954. The Laureates are required to deliver a lecture and provide a written version for publication, typically somewhat expanded. Ahmed's published Nobel Lecture was extraordinary.[15] It was book-length, with three dozen figures, many of striking artistic quality. It presents, in lucid and often elegant exposition, a fabulous scientific odyssey that elucidates history as well as experiments and concepts. He also highlighted the contributions of many others, within his group and beyond.

In an epilogue, Ahmed said: "Personally, I did not originally expect the rich blossoming in all of the directions outlined in this anthology." Then he described the thrill of discovery, quoting Howard Carter's amazement at his first glimpse into Tutankhamen's Tomb: "At first, I could see nothing ... then shapes gradually began to emerge .... Beautiful things."

Ahmed visited Harvard again in May 2001 to give the Kistiakowsky Lecture. His talk, titled "The Femtosecond Realm: Exploring New Limits," was a wonderfest. He described advances and prospects for femtobiology involving electron transfer in DNA and proteins, ultrafast electron diffraction, and laser-induced control of reaction pathways (but not yet for any of the reactions on my ballot). Some of his slides illustrated the classic stop-motion photography of a trotting horse, precursor of movies and kindred to his laser femtochemistry. Those slides came from a comprehensive paper soon after that was published in the *J. Chem. Edu.*,[16] titled "Freezing Atoms in Motion; Principles of Femtochemistry and Demonstration by Laser Stoboscopy." It was designed for high-school students! Basic laser apparatus was displayed and particle-wave duality neatly explained, along with the concepts of wave packets and coherence. His lecture also addressed why the Heisenberg uncertainty principle was not an obstacle to obtaining femtosecond resolution. Robust coherence was the key, as Ahmed soon explained again in a lyrical one-page commentary in *Nature*,[17] titled "The Fog That Was Not."

In our conversations, Ahmed wanted to hear stories about both George Kistiakowsky and Bright Wilson. Close friends for 50 years, they both were revered as scientists and citizens.[18,19] Here I will just mention a pair of Kisty stories that Ahmed liked. In the early 1950s, Kisty undertook a molecular beam experiment, but it fizzled because he did not have adequate vacuum pumping. Frustrated, he smashed the apparatus with an axe! Less than a decade later, when he asked me what research I was going to pursue, Kisty cautioned: "Too bad! You have been bitten by the crossed molecular beam bug! There are hardly any collisions in one beam, so there are still fewer when crossing two beams!" Less than another decade later, I told him that "the splinters from his axe sprouted into a beautiful garden."

Ahmed knew extensively about Kisty's contribution to science and public service but was interested to hear more about his work with the Council for a Livable World, a lobby for peace. Describing the Council, on which I had long been a member, led to speaking about its founder, Leo Szilard,[20] and his allegorical fable, *The Voice of the Dolphins*, published in 1962. It presaged the amazingly bloodless political transformation of Eastern Europe in 1989. We discussed the role of TV, which showcased the Western world to people behind the Iron Curtain. That generated a collective awareness that served as the "Voice of the Dolphins." Ahmed then told me about another commentary he had just published in *Nature*,[21] titled "Science for the Have-nots," based on his recent address to the Pontifical Academy of Science at the Vatican. After a somber assessment of barriers, Ahmed stressed essential ingredients for progress. Among them he urged "ensuring active participation of women in society … reforming education … creating a credible legal code … and building a science base."

Two years later, he published a major manifesto,[22] *The Future of Our World*, in which Ahmed called for a rational world vision and a universal perspective that would be unifying for humanity: "The key is not to ignore the have-nots, not to ignore the frustrated part of the world, politically and economically, and to recognize that poverty and hopelessness are the primary sources of terrorism and the disruption of world order."

During his 2001 visit, Ahmed spoke of the proposal he had made to create a University of Science near Cairo to arouse the imagination of Egyptians and demonstrate the power of science in building the future. As he had hoped, its cornerstone was laid on January 1, 2000. After a decade of delays, the project was revived in 2011, following the January 25 Revolution. Now named the Zewail City of Science and Technology, construction is well underway. On its website, Ahmed posted a *Founders Statement*. He described the City's five components and "the path from university education to research and development and on to the local and global economy market." He added: "Just as the Aswan Dam provided the power to build the industrial base … in this age [we] must provide power for the mind, as *Knowledge is the Light of Life* … a knowledge-based economy is the only way to improve national productivity." Gratefully, he wrote that "Egyptians inside and outside the country are continuing to make donations to the City … I am confident that we will succeed and prove the City's motto *Egypt Can* during this historic epoch of Egypt and Arab Awakening."

Ahmed anticipated that the first phase of the new campus would be inaugurated before Fall 2016. He died shortly before. He had arranged to be buried near the City. A dear friend, Mostafa El-Sayed, said: "He was hoping to keep reminding the Egyptian people of the importance of scientific research so that Egypt can return to its ancient technical and scientific glory".[23]

In his essay on Franklin, the concluding words by Ahmed are fitting as an epitaph for himself: "[His] vision has many dimensions that together formed the legacy of a great man in search of a better world ... his life's work remains as a monumental book of knowledge ... [that] shall continue to be a beacon of enlightenment."

References

1. A.H. Zewail, "Optical Molecular Dephasing: Principles of and Probings by Coherent Laser Spectroscopy," *Acc. Chem. Res.* **13**, 360 (1980).
2. N.F. Scherer, J.L. Knee, D.D. Smith, and A.H. Zewail, "Photofragment Spectroscopy: The Reaction ICN → CN + I," *J. Phys. Chem.* **89**, 5141 (1985).
3. R.B. Bernstein, D.R. Herschbach, and R.D. Levine, "Dynamical Aspects of Stereochemistry," *J. Phys Chem.* **91**, 5465 (1987).
4. S. Buelow, G. Radhakrishnan, J. Catanzarite, and C. Wittig, "The Use of van der Waals Forces to Orient Chemical Reactants: The H + CO_2 Reaction," *J. Chem. Phys.* **83**, 444 (1985); S.I. Ionov, G.A. Brucker, C. Jaques, L. Valachovic, and C. Wittig, "Subpicosecond Resolution Studies of the H + CO_2 → CO + OH Reaction Photoinitiated in CO_2-HI Complexes," *J. Chem. Phys.* **99**, 6553 (1993).
5. N.F. Scherer, L.R. Khundkar, R.B. Bernsein, and A.H. Zewail, "Real-time Picosecond Clocking of the Collision Complex in a Bimolecular Reaction: The Birth of OH from H + CO_2," *J. Chem. Phys.* **87**, 1451 (1987).
6. A.H. Zewail and R.B. Bernstein, "Real-Time Laser Femtochemistry: Viewing the Transition States from Reagents to Products," *Chem. Eng. News* **66**, 24 (1988).
7. A.H. Zewail (Ed.) *The Chemical Bond: Structure and Dynamics* (Academic Press, Boston, 1992).
8. "Memorial Issue for Richard B. Bernstein," *J. Phys. Chem.* **95**, 7964–8421 (1991); also, J.L. Kinsey and R.D. Levine, *Biographical Memoir for R.B. Bernstein* (National Academy of Sciences, Washington, DC, 1998).
9. I.R. Sims, M. Gruebele, E.D. Potter, and A.H. Zewail, "Femtosecond Real-Time Probing of Reactions. VIII. The Bimolecular Reaction of Br + I_2," *J. Chem. Phys.* **97**, 4127 (1992).
10. S.I. Chapin, "A Legendary Bon Mot?: Franklin's 'What is the Good of a Newborn Baby?'" *Proc. Am. Philos. Soc.* **129**, 278–290 (1985); I.B. Cohen, "Faraday and Franklin's 'Newborn Baby'," *Ibid.*, **131**, 177–182 (1987).

11. A.H. Zewail, "Franklin's Vision," *Proc. Am. Philos. Soc.* **150**, 542–552 (2006).
12. Q. Liu and A.H. Zewail, "Femtosecond Real-Time Probing of Reactions: Reaction Times and Model Potentials," *J. Phys. Chem.* **97**, 2209 (1993).
13. A.H. Zewail, "Femtochemistry," *J. Phys. Chem.* **97**, 12427 (1993).
14. D. Herschbach, "Linus Pauling: Quintessential Chemist," in *Pauling's Legacy: Modern Modeling of the Chemical Bond*, Z.B. Maksic and W. J. Orville-Thomas (Eds.) (Elsevier Science, Amsterdam, 1999).
15. A.H. Zewail, "Femtochemistry: Atomic Scale Dynamics of the Chemical Bond using Ultrafast Lasers," in *Les Prix Nobel, The Nobel Prizes 1999* (Norstedts Truckeri AB, Stockholm, 2000). Also, a considerably shortened version is A.H. Zewail, *J. Phys. Chem. A* **104**, 5660–5694 (2000).
16. J.S. Baskin and A.H. Zewail, "Freezing Atoms in Motion: Principles of Femtochemistry and Demonstration by Laser Stroboscopy," *J. Chem. Ed.* **78**, 737 (2001).
17. A.H. Zewail, "The Fog That Was Not," *Nature* **412**, 279 (2001).
18. F. Dainton and G.B. Kistiakowsky (1900–1982), *Biogr. Mem. Fellows R. Soc.* **31**, 338–376 (1985).
19. D. Herschbach, and E. Bright Wilson (1908–1992), *Nucleus* **72**(7) (1994).
20. W. Lanouette, *Genius in the Shadows* (Macmillan Publishing, New York, 1992); L. Szilard (1898–1964), *The Voice of the Dolphins and Other Stories*, New Edn. (Stanford University Press, 1992); D.R. Herschbach, "The Dolphin Oracle," *Harvard Magazine* **95**(57) (1993).
21. A.H. Zewail, "Science for the Have-Nots," *Nature* **410**, 741 (2001).
22. A.H. Zewail, "The Future of Our World," in *Einstein … Peace Now! Visions and Ideas*, R. Braun and D.K. (Eds.) (Wiley-VCH, Amsterdam, 2005), pp. 109–123.
23. M.A. El-Sayed, "Ahmed Zewail: A Force for Egyptian Science," *Proc. Natl. Acad. Sci.*, **114**, 1743 (2017).

Author Biography

Dudley Herschbach was born in San Jose, California, in 1932, and grew up nearby in a rural farming area (pre-Silicon Valley!). He did not expect to go to college, much less become a professor and scientist! He owes his career to inspiring teachers and generous scholarship awards. At Stanford University, he received his B.S. in Mathematics in 1954, mentored by George Polya, and his M.S. in Chemistry in 1955, mentored by Harold

Johnston. At Harvard, he obtained an M.S. degree in Physics in 1956, mentored by Robert Pound, and his Ph.D. in Chemical Physics in 1958, mentored by E. Bright Wilson. Dr. Herschbach then joined the chemistry faculty at University of California, Berkeley, in 1959, and undertook molecular beam experiments to elucidate the dynamics of chemical reactions in single-collisions. He returned to Harvard in 1963, expanding upon beam research and much else over the next four decades. The research thrived, attracting graduate students and postdocs of exceptional ability and adventurous spirit. His teaching spanned widely over graduate and undergraduate courses, including freshman chemistry for 20 years, which was his most challenging assignment. Dr Herschbach has served as an Emeritus at Harvard since 2003 but continued teaching a freshman seminar (up to 2011, titled "Molecular Motors: Wizards of the Nanoworld"). Since 2005, he is an itinerant (i.e., part-time) member of the physics faculty at Texas A&M University. His evangelical efforts to enhance K-12 science education centered for many years on high school Science Fairs (see *The Archimedes Initiative* website). In addition, he has given many lectures, interviews, and radio and TV appearances, including as a guest voice on *The Simpsons*. Dr. Herschbach served on boards of the Society for Science and the Public and Council for a Livable World. His honors include the Nobel Prize (1986), shared with Yuan Lee and John Polanyi; among others much appreciated are Honorary Life member of the Association of Women Scientists (1998); and three awards named after him: the Harvard Chemistry Teaching Prize for Graduate Students (annual since 2003), the Harvard University Teacher/Scientist Lectureship (annual since 2006), and the Molecular Dynamics Prize (biannual since 2007).

Chapter 2

Ahmed Zewail: Larger than Life

*Roger Kornberg**

Ahmed Zewail was a larger-than-life personality. He had a commanding presence, which cannot easily be put in words. Like anyone or anything great, there is no substitute for the original. His love of science, love of people, and love of life only begin to describe the totality of the amazing man. I am reminded of the lines from Shakespeare:

> *His life was gentle, and the elements*
> *So mixed in him that Nature might stand up*
> *And say to all the world, "This was a man."*

I met Ahmed when he was already accomplished and widely revered, and I was in every respect his junior. Yet I was immediately swept up in his literal and figurative embrace. He encouraged me, promoted my fortunes, and was from the first moment the truest of friends. I have never known anyone so warm and genuine in this way. I cherish the memory.

---

*kornberg@stanford.edu

Author Biography

Roger Kornberg received his B.A. degree from Harvard College in 1967 and Ph.D. in Chemistry from Stanford University in 1972. He is the Winzer Professor in Medicine in the Department of Structural Biology at Stanford University. He was a postdoctoral fellow and member of the scientific staff at the Laboratory of Molecular Biology in Cambridge, England from 1972 to 1975, and a faculty member of the Department of Biological Chemistry, Harvard Medical School, from 1976 to 1977. He moved to his present position in 1978, where his research has focused on the mechanism and regulation of eukaryotic gene transcription. Kornberg was elected to the National Academy of Sciences in 1993. He has received many awards, including the Welch Prize (2001), highest award in chemistry in the United States; the Leopold Mayer Prize (2002), highest award in biomedical sciences of the French Academy of Sciences; and the Nobel Prize in Chemistry (unshared, 2006). Kornberg's closest collaborator has been his wife, Dr. Yahli Lorch. They have three children, Guy, Maya, and Gil.

Chapter 3

Stories from the Round Table

Marshall Cohen*, Charlie Campbell†
and Rudolph A. Marcus‡

Ahmed was a member of one of the famous round tables at Caltech's Athenaeum, where faculty would meet for lunch and discuss the events of the day, debate recent scientific results, and share an occasional joke. For his memorial on January 19, 2017 (appropriately held at the Athenaeum), his lunch companions were asked to speak about Ahmed and tell stories from his days at the table. The following are the texts of the speeches given that day.

> *Marshall Cohen is Professor Emeritus of Astronomy at Caltech. He is a radio astronomer from the days when, to be a radio astronomer, you had to first be an electrical engineer. He was first from the table to speak:*

I am from the Astronomy Department at Caltech, not Chemistry, and did not know Ahmed from his work there. No, I knew Ahmed from the Round Table at the Athenaeum. Actually, as many of you know, there are several of these faculty tables, and there is a strong tendency for participants to sit at the same table day after day, and even to occupy the same seat. Ahmed

---

* mhc@astro.caltech.edu
† campbell@usc.edu
‡ ram@caltech.edu

and I usually sat at what is sometimes called the Old Folks Table, perhaps because the median age there is in the 80s. Currently it is around 86. I am close to the median, from above. Ahmed was below.

We used to have a lot of chemists at the Table. There was Jack Roberts, Jack Richards, Nelson Leonard, and Ahmed Zewail, and sometimes Aron Kupperman, and now only Rudolph A. Marcus is left. Of course, Rudolph counts for two, but we need more chemists.

The table was always livelier when Ahmed was there. Even when we were talking about soccer, or electric cars, or stellar evolution, his enthusiasm, and smile, and questions would keep us going. He was the center of attention. I remember very well several occasions when he came in, all smiles and evidently holding in a secret, and he let us in on some new result, and how it would be in *Nature* soon. He loved his work.

Ahmed was proud of his work in Egypt, of the fact that streets and schools there were named after him, of the stamps that carried his picture.

Ahmed needed to understand how things worked. He questioned me closely about radio interferometry, my own area, about the coherence and why the system was built in this way or that. He had an opinion about most things, or at least a question that usually needed a thoughtful answer. He did defer to Francis Clauser, an aerodynamicist, and for many years, the Elder of the Table, who had studied hieroglyphics and had a remarkable knowledge of Pharonic Egypt.

Ahmed and I would often walk back down the Olive walk, to our respective buildings; he to Noyes and I to Robinson (later, to Cahill across California Boulevard). We would continue the noon-time discussion, but, when it was just the two of us, we could be more intimate. Sometimes we talked about religion. I the secular Jew and he the secular Moslem agreed that religion was an important factor in the lives of many people, and should not be denigrated. He talked about his mother, a devout woman for whom religion was very important. I tried to describe some of the rules followed by orthodox Jews, and why they are important in that society.

Ahmed was a great scientist and teacher, and a good friend. We all miss him.

Charlie Campbell is a Professor of Aerospace and Mechanical Engineering from down the road at USC, who joined the round table during a long-term visiting appointment at Caltech and never left. He remembers:

"A cappuccino...extra hot!"

This was how Ahmed would complete his lunch every day, shouting out the order while pointing to the ceiling. And he relished the coffee. I remember in particular that he used to take the raw sugar cubes that come with the coffee, dip then in the foam and then touch it to his lips and with a big smile suck out the juice. Now I can tell you, I've also ordered "A cappuccino...extra hot!" (I even tried pointing to the ceiling as I ordered) and, honestly, I don't think it was hotter than any other cappuccino, but I'm sure it made Ahmed feel like he was getting the special treatment he so justly deserved.

Along with Marshall, I had the privilege of eating lunch with Ahmed for nearly 10 years at the famous round table. Because of that, I was asked to say a few words about it. The table seats nine, easily expandable to ten, and occasionally to eleven, which is just about perfect for an inclusive discussion. The table was presided over by Francis Clauser and in the early days (or at least my early days) included as regulars or frequent visitors Ahmed, myself, Nelson Leonard, Ned Munger, Bob Christy, Charlie Barnes, Jack Roberts, Jack Richards, Lee Silver, Joe Kirschvink, Maarten Schmidt, Marshall Cohen, Rudolph A. Marcus, David Baltimore, Alice Huang, Ed Stopler, and Kerry Sieh (sadly too many of these people are no longer with us). And then there was the occasional outside visitor who could expect to be quizzed and hounded until every scrap of information had been drawn from him. This was a real Bloomsbury group. The attraction of the table was great conversation between intelligent well-informed friends. The topics were wide-ranging, from recent scientific advances (for which there was nearly always an expert available), history, and of course politics. An almost constant topic was foreign affairs, and Ahmed was our expert in the Middle East and the Muslim world, which after 9/11 became quite a job. Francis Clauser usually led the debate. Francis was quite a fellow and could argue both sides of almost any issue. He would weave the conversation around him like a tapestry. Often he would turn the conversation by pointing at one of us and shouting "You're dead wrong" — that is if you can imagine that being said in the friendliest possible way. As Ahmed would sit directly across the table from Francis, he was by default the dead-wrongest of us all. But he would often respond with his wide smile and boisterous laugh.

Ahmed had other little quirks. He was fond of carefully scanning the menu while mumbling "Now what here has zero calories?" He was fond of reminding Bob Christy, who was President when Ahmed was promoted to Associate Professor with tenure, that his promotion did not come with an increase in pay. But that was all done in good humor (but as he told it over and over again — maybe not entirely in good humor). And finally he could make use of the resources at the table which were wide and varied. As Marshall mentioned, Francis Clauser, for the grand purpose of why-the-hell-not, had learned to read and write in hieroglyphs, and so it was Francis who translated Ahmed's Order of the Nile collar.

Over the years you learn a lot about a friend. There are things to be proud of: Ahmed deeply regretted that the Muslim world had fallen from the pinnacle of a thousand years ago when it lead the world in science and almost any other intellectual pursuit. At lunch, he pursued and debated the historical reasons behind the fall. He strongly desired to make it once again a center of science. You can hear this in numerous op-ed pieces that he wrote. In 2000, he laid the cornerstone for what was to become the Zewail City. When KAUST (the King Abdullah University for Science and Technology) opened in 2009, he was on the board. In 2011, he was in Cairo for the Tahrir Square protests during which he was alternately felt revered or feared for his life. Finally, first under Mubarak and then under Sisi, he was able to bring Zewail City into fruition.

And there are some that are not so good: Ahmed voted for George W. Bush in 2000 and quickly regretted it. I think he had too much faith that democracy would ultimately do the right thing and that he voted for Bush to free us from the scandals and sexual malfeasance of the Clinton administration. But he was soon appalled by Bush's fall off the cliff into war and failed international relations. Ahmed, of course, was vehemently against the Iraq war. I remember him spotting me following him on the path to the Athenaeum a few months before the start of the Iraq war, stopping and turning around, literally shaking his head and wringing his hands in frustration, saying "what are we going to do to stop this man?" Sadly we couldn't.

Ahmed attended the table nearly every day until 2010 when he just became too busy particularly after being appointed by President Obama to PCAST (Presidential Council of Advisors on Science and Technology), and his appointment to be the US Scientific Ambassador to the Middle East. That on top of KAUST, Tahrir Square, and Zewail City — and eventually his illness — left little time for us. For the last few years, we were lucky to see him once or twice a year (usually on a birthday, when there was cake).

And sadly that epitomizes the round table culture here at Caltech. Twenty years ago, there were four round tables and you had to show up early to get a seat at the Clauser/Zewail table. But for whatever reason, that culture is disappearing as the senior members that have passed are not replaced by younger faculty. Now it is easy to find five empty seats at the former Clauser table. But, if you talk to the those that grew up in that culture, they will tell you the advantages that come from being scientifically stimulated by lunch discussions, sometimes by direct scientific discussion, but more often by just being intellectually stimulated by non-scientific discussion. After all, science proceeds in incremental steps, but genius...genius comes from that rare out-of-the-box moment brought on by — who knows what? Wherever it comes from, it isn't from the next step in a derivation or, for that matter, from anything within whatever field you are studying. Sometimes it comes from an astrophysicist explaining astrophysics to an engineer, or an engineer explaining engineering to an astrophysicist, or maybe during an idle moment of a political discussion. But it ain't gonna happen if there have no outside stimulation and for this the round tables are invaluable. Ahmed certainly knew this. So young-folk, don't so easily pass up your opportunities and come join us.

Let me close by mentioning my favorite picture from his autobiography, one of Ahmed taking Hani and Nabeel trick-or-treating, with Ahmed wearing the sumptuous robes from his Oxford honorary degree. As Ahmed told the story, a neighbor asked him where he had bought his costume and Ahmed, always truthful, replied "Target."

Rudolph A. Marcus is John G. Kirkwood and Arthur A. Noyes Professor of Chemistry at Caltech and is in sole possession of the 1992 Nobel Prize in Chemistry. He was in Singapore during Ahmed's memorial, but sent the following short message for Charlie Campbell to read in his stead.

Dear Charlie,

I am sorry that I will miss your presentation today that perhaps will include life with Ahmed at the round table. I am here in Singapore, interacting with some 100+ high school students from a variety of countries (9th International Science Youth Forum). Had I been at the memorial in honor of Ahmed today, I certainly would have commented on the utter delight of our countless walks between Noyes and the Ath on our way to and from the round table lunches in those pre-Nobel halcyon days and on the influence on our research.

All good wishes and regards,

Rudolph

> *For this submission Rudolph added the following "...some of the remarks I would have made had I been at the memorial:"*

I first met Ahmed when he visited the University of Illinois in 1975 and I was chairman of the Chemistry Staffing Committee. As kindred spirits, we enthusiastically "hit it off," our age difference of some 23 years notwithstanding. The department was looking for someone involved in molecular beams, lasers, and chemical reactions. It was almost a match. Ahmed's interests were in molecular beams, lasers, and dephasing of coherence. In any case, the department offered the position to someone else, and Ahmed returned to Berkeley for another year postdoctoral with Charles Harris, and in 1976 came to Caltech. Clearly, Ahmed did not bear a grudge. Harry Gray as the next department chairman and Ahmed as a vigorous young chemical physicist searching for new faculty were both influential in my coming to Caltech from Illinois in 1978.

Ahmed would not have minded my telling this story, since he was very fond of it himself and its irony. The next 25 years or so were ones of glorious camaraderie and discussions, many of them on our way to and from the Ath round table. Initially, in some of these discussions, he tried to persuade me to get back to theory involved with real experiments and

I tried to persuade him to move from coherence to reactions. In any case, both changes occurred and probably would have occurred without the discussions. Our offices at Caltech are next door to each other and often I would hear his booming and enthusiastic voice that radiated warmth. Sadly, the hall is now quiet.

Chapter 4

Ahmed Zewail: Great Scientist, Dear Friend

*Harry B. Gray**

I met Ahmed Zewail in the 1970s, during the time I was Chair of the Caltech Division of Chemistry and Chemical Engineering (CCE) Staffing Committee. Ahmed had applied for the faculty opening in experimental chemical physics, and I was very interested in him. I was impressed with the graduate work he had done at Penn with Robin Hochstrasser as well as his postdoc research at Berkeley with Charles Harris. I called Charles, a longtime friend, who told me to go all out to attract Ahmed to Caltech. I did!!

    I invited Ahmed to Caltech for an interview. His talk was terrific, maybe a little too good, as some of the more conservative faculty members were not convinced he was for real. We decided to postpone our faculty search for a year, and during that time, Ahmed received offers from many top research universities. He and I had many phone conversations during that year, and I pleaded with him to wait until we could get our act together and make him an offer. We bonded during that time, and I brought him back for a second visit. Very fortunately, by then our "conservatives" had come to their senses and supported the appointment with

---

*hbgray@caltech.edu

great enthusiasm. We made the offer, Ahmed accepted, and he moved from Berkeley to Caltech in 1976, shortly before I began my term as CCE Chair.

Within a few months, Ahmed was running a talented and dedicated research group, designing and constructing instruments for investigations of the earliest events in chemical reactions. Very soon he and his team were publishing groundbreaking papers in the best journals, and by the mid-1980s, their work was making waves nationally and internationally. At about this time, to my great surprise, I discovered that Ahmed had not won any national awards. My colleagues and I decided to fix this oversight, and during our discussions, Bill Goddard pointed out that the Buck–Whitney Award of the Eastern New York Section of the American Chemical Society, which he had won in 1978, was a prize for "outstanding contributions to pure and applied chemistry by someone who has not yet achieved national recognition." I nominated Ahmed immediately for the Buck–Whitney, and he won the award in 1985, just before I stepped down as CCE Chair.

As a Nobelist 14 years later, Ahmed made the Buck–Whitney famous! Indeed, the award from the Eastern New York Section was the very first entry in a seemingly endless parade of honors to my friend and colleague! I take great pride not only in recruiting him to Caltech but also in taking the lead that led to his first award.

I had many enjoyable discussions with Ahmed about joint interests in science, most especially when he turned his attention to ultrafast processes involving transition metal complexes. I well remember the excitement we shared when he told me he had determined the rates of both (phototriggered) Mn–Mn and Mn–CO dissociation in dimanganese decacarbonyl in the gas phase. After ultrafast laser excitation, the Mn–Mn single bond ruptured in a few femtoseconds, generating two pentacarbonyl fragments, as predicted from a calculated triplet excited state energy surface based on earlier work I had done at Columbia with Bob Levenson; the axial CO came off shortly thereafter, a bit slower, as expected, owing to the relatively massive product fragment, dimanganese nonacarbonyl.

Our last discussions regarding science took place in his conference room during his struggle with cancer. He and a postdoc, Byung-Kuk Yoo, were working on a paper on titanosilicate catalysis in collaboration with my good friend Sir John Meurig Thomas. I had written many papers on the electronic structures of metal–oxo complexes, and Ahmed thought I could be of help with the interpretation. In the days that followed, we had many good discussions of titanium-oxo structure and reactivity.

In their *PNAS* paper, Ahmed and collaborators demonstrated that a multiply bonded Ti(IV)oxo unit in a low-coordinate titanosilicate catalytic center could be activated by laser-induced electron transfer to form a stretched (singly bonded) Ti(III)oxo. The femtosecond dynamics they elucidated featured exceptional Ti–O bond disruption, making the oxo very basic for interaction with a proton on water or on the carbon of carbon dioxide (or carbon monoxide). This groundbreaking work has given us (for the first time!) snapshots of early events in water splitting (as well as carbon dioxide reduction) catalytic cycles!

Rest in peace, Ahmed. You will be remembered as long as science is pursued on Planet Earth.

Author Biography

Harry B. Gray is the Arnold O. Beckman Professor of Chemistry and the Founding Director of the Beckman Institute at the California Institute of Technology. After pursuing graduate work in inorganic chemistry at Northwestern University and postdoctoral research at the University of Copenhagen, he joined the chemistry faculty at Columbia University, where in the early 1960s, he investigated inorganic electronic structures and reactions. At Caltech since 1966, he has worked on electron transfer in proteins and on inorganic chemistry with emphasis on the development of catalysts for the production of solar fuels. Among other honors, he has received the National Medal of Science (1986); six national awards from the American Chemical Society, including the Priestley Medal (1991), the Linderstrom–Lang Prize (1992), the Harvey Prize (2000), the Wolf Prize in Chemistry (2004), the Welch Award in Chemistry (2009), and the Othmer Gold Medal (2013); and 19 honorary doctorates. He is a member of the National Academy of Sciences, the American Academy of Arts and Sciences, and the American Philosophical Society, and a foreign member of the Royal Danish Academy of Sciences and Letters, the Royal Swedish Academy of Sciences, the Royal Society of Great Britain, and the Accademia Nazionale dei Lincei.

Chapter 5

In Memory: Ahmed Zewail

Jack Dunitz*

It was in March 2013 during a visit to Zurich to give the Richard Ernst Lecture that the first signs of Ahmed's fatal illness brought him to the Zurich University Hospital and kept him from attending a dinner party arranged by Richard Ernst at his home in Winterthur. Naturally, we were alarmed by the news of this medical emergency and disappointed by his absence. I never saw Ahmed again. We had hoped that subsequent treatment in Pasadena would lead to rapid and complete recovery. But it was not to be.

My first meeting with Ahmed was during a visit to Caltech in the mid-1980s, a few years after his appointment as a young faculty member. I was immediately impressed by the scope and grandeur of his plans, by what he hoped to achieve — following the course of chemical reactions at the femtosecond level, ultrafast pulsed electron diffraction, and microscopy — although I must admit to having had doubts as to whether he could ever succeed in these grandiose projects. Here was a junior faculty member with a program that seemed appropriate for an entire Research Institute!

Ahmed and I soon found that we shared an admiration for the life and work of Linus Pauling, whom we both regarded as one of the greatest scientists of the century. I was touched by Ahmed's desire to heal the rift

---

*jack.dunitz@org.chem.ethz.ch

that had occurred between Caltech and Pauling 20 years earlier, when, following the award of the Nobel Peace Prize, Pauling had left Caltech with bad feelings on both sides. Pauling's activities directed at the banning of atmospheric atom bomb tests had not found general approbation among the Caltech leadership and even among some of his scientific colleagues. While his first Nobel Prize in 1953 had been greeted with pride and enthusiasm by Caltech as a whole and by the local press, his second Prize, nine years later, was met with silence or even outspoken disapproval. Disappointed, Pauling had left the institute which he had established as one of the leading chemistry departments in the world. There had not been enough effort from his colleagues to restrain him from leaving. An additional factor for Pauling's departure may have been his developing interest in "orthomolecular medicine." In any case, whatever the underlying causes, there had been a rift between Caltech and Pauling, and Ahmed felt that it was his business to restore good relations between Caltech and the man who had made it one of the major chemistry research centers in the world. With tact and patience he succeeded. He established a close personal relationship with his illustrious predecessor and went on to organize symposia at Caltech to celebrate Pauling's 85th and 90th birthdays. Later, after Pauling's death, Ahmed organized a special meeting at Caltech in 2001 to mark the centenary of Pauling's birth, and I feel privileged to have been one of the speakers at this event. It was a fitting mark of appreciation by Caltech to appoint Ahmed as first Linus Pauling Professor of Physical Chemistry.

I remember that soon after we met I asked Ahmed if, as an Egyptian scientist, he would have any problems about accepting an invitation to visit the Weizmann Institute. He told me he had no qualms about visiting Israel and would be glad to accept such an invitation. Next time I was in Israel, I passed this information on to my Weizmann friends, and it was not long before Ahmed was invited to Israel. Of course, I am well aware that others may also have encouraged such a visit. I was enormously pleased when Ahmed was awarded the Wolf Prize in 1993, one of the earliest major prizes he received in recognition of his remarkable achievements, which were also honored in Zurich by the award of the Paul Karrer Gold Medal in 1998, just a year before his Nobel Prize. I mention these awards because, in my opinion, awards given to a Nobel Prize winner prior to the visit to Stockholm count far more than those that come later. They point the way rather than follow the trend. With his achievements, abilities, and wide interests, Ahmed became an important scientific advisor to his native country Egypt as well as to his adopted home of USA. In addition to his efforts to advance science and general education in Egypt, he was a member of President Obama's Scientific Advice Council and Science Advisor to the United Nations.

As the years passed, my visits to Caltech became less frequent and then came to a stop, but my contact with Ahmed was maintained to some extent through the medium of my old friend Sir John Meurig Thomas, who has been collaborating with Ahmed for many years in the development of what has come to be known as 4D Electron Microscopy. Together they wrote a book on the achievements and possibilities of this new technique that can provide a sequence of three-dimensional images at atomic level and at femtosecond time intervals. John and Ahmed became close friends, and I have had the benefit of regular reports on Ahmed's activities over recent years through regular reports from John.

I count myself fortunate to have met Ahmed Zewail and to retain in my memory fragments of our encounters over the years. I shall miss his smile, his good sense, his optimism, his enthusiasm, and his ability to find the right word at the right time.

Author Biography

Jack Dunitz (b. 1923) studied chemistry at Glasgow University and earned his Ph.D. in 1947 and held research fellowship at Oxford University (1946–1948, 1951–1953), California Institute of Technology (1948–1951, 1953–1954), US National Institute of Health, Bethesda, MD (1954–1955), and the Royal Institution, London (1956–1957), before taking up a professorship at the ETH-Zürich, a post he held until his retirement in 1990. He has held Visiting Professorships in the US, Israel, Japan, Canada, Spain, and the United Kingdom. Dunitz's research involved the use of crystal structure analysis as tool for studying chemical problems, e.g., structure and reactivity of medium-ring compounds, ion-specificity of natural and synthetic ionophores, chemical reaction paths, molecular motions, polymorphism, solid–solid phase transformations, and solid-state chemical reactions. He was a recipient of the Paracelsus Prize of the Swiss Chemical Society (1986), the Gregori Aminoff Prize of the Royal Swedish Academy of Sciences (1990), and the Buerger Award of the American Crystallographic Association (1991), among others. Dunitz is a Fellow of the Royal Society of London and member of the German Academy of Sciences Leopoldina, the Academia Europaea, and the European Academy of Sciences and Arts. He is a foreign member of the Royal Netherlands Academy of Sciences, the US National Academy of Sciences, the American Philosophical Society, the American Academy of Arts and Sciences, and also Honorary Member of the Swiss Society of Crystallography, British Crystallogaphic Association, Royal Society of Chemistry, and Swiss Chemical Society. Since his official retirement, he has worked on problems of polymorphism, phase transformations in solids, weak intermolecular interactions in condensed phases, and crystal structure prediction. Dunitz has written more than 350 scientific papers and is the author of the books *X-Ray Analysis and the Structure of Organic Molecules* (Cornell University Press, 1979; Verlag HCA, Basel, 1995) and *Reflections on Symmetry in Chemistry ... and Elsewhere* (with E. Heilbronner, Verlag HCA, Basel, 1993).

Chapter 6

Some Personal Recollections of Ahmed Zewail

*Fred C. Anson**

I did not know Ahmed Zewail (AZ) before he accepted a faculty position in Chemistry at Caltech. But shortly after he joined the chemistry faculty, I became Chairman (of the Division of Chemistry and Chemical Engineering) and he and I soon became well-acquainted. I was an experimental electrochemist, not a chemical physicist like AZ, but it wasn't long before our interactions resulted in our becoming good friends. (And not only because my appointment as Chairman put me in position to provide some support to AZ's research activities!)

We worked together to create an "Industrial Consortium," which provided considerable research support to many members of the faculty for a number of years.

In addition, he and I organized two symposia/banquets in celebration of Linus Pauling's 85th and 90th birthdays. Pauling attended both events with enthusiasm and recalled in his after-dinner remarks that it was at Caltech where he had conducted the research that lead to his receiving his first Nobel Prize (in Chemistry). He noted that an important enabling feature throughout his years at Caltech was the presence of colleagues such

---

*fanson@caltech.edu

as AZ who stimulated and appreciated his research. Pauling was quite delighted when AZ was named the first Linus Pauling Professor of Chemistry.

Among the many admirable attributes exhibited by AZ was his enthusiasm for modern chemistry, including the challenges it presented to those who wished to expand our understanding of the intricate details of chemical reactions. Equally evident and praiseworthy was his devotion to his students and post-doctoral associates. The eminent professional positions that so many of them have attained are just one monument to AZ's inspiring mentorship.

Another engaging aspect of AZ's personality was his sense of humor. He often pointed out the ironic or amusing features of research activities, including those of some of the scientists who pursued them. The preparation of the joint research papers that he and I published was made more enjoyable by the, often humorous, comments he offered to our mutual efforts. He was particularly proud of our paper that was featured on the cover of *Angewandte Chemie*.

AZ's deep appreciation of and commitment to his Egyptian heritage, especially to enhance the education of Egyptian students, were admirably evident throughout his lifetime. His commitment and effective activities to foster this goal were certainly achievements in which he took particular pride. Science education in Egypt will benefit from AZ's influence and interest in the years ahead.

Chemical physics at Caltech and throughout the world were profoundly advanced by AZ. It was my surpassing pleasure to have assisted a little in his accomplishments. My memories of our association are full of respect, admiration, and happiness.

Author Biography

Fred C. Anson is Elizabeth W. Gilloon Professor of Chemistry Emeritus at the California Institute of Technology. He received a B.S. in Chemistry from Caltech in 1954, and a Ph.D. from Harvard in 1957. Upon receiving his Ph.D. degree, Dr. Anson returned to Caltech, becoming a member of the faculty in the Division of Chemistry and Chemical Engineering.

He later served as Chairman of the Division (1984–1994), and became Professor Emeritus in 2001. Dr. Anson is a Fellow of the John Simon Guggenheim Foundation and the Alfred P. Sloan Foundation. He was elected to the National Academy of Sciences in 1986. He received a D.H.C. from the University of Paris in 1993.

Chapter 7
My Friend Ahmed and I

Bengt Nordén*

Ahmed and I were close — at least that is what I always felt — he called me his big brother (I was one year his senior). While I was more direct about most things, even personal matters, he was generally more difficult to read and, everybody I know agrees, intrinsically quite private. However, we shared the same sense of humor and the same curious, I would say almost childlike, mindset when approaching scientific problems — though his was that of a genius. Ahmed was, despite his deeper felt reservation, very social, with an eye for everybody, whether a janitor or a Nobel laureate. To really see people is something I think I learnt very much from him and indeed find very important in life. His social ability was always combined with humor: I remember many a good laugh together. At the same time, he was reflective and a deep philosopher — often finding scholarly parallels with ancient Egyptian or Greek science.

We met more frequently after his Nobel Prize win, when our families had got to know each other in Stockholm. Actually, the first time we met was only some five years before the prize was awarded to him. Ahmed was touring in Sweden, giving seminars at the universities and the story, as Ahmed loved to tell it, goes as follows. My friend and former colleague from my alma mater University of Lund, then Chair Professor of Inorganic

---

* norden@chalmers.se

Chemistry at KTH in Stockholm, Ingmar Grenthe, one day called me and suggested I host Ahmed Zewail for a seminar at Chalmers. I must have asked "Zewail who?" and Ingmar, quite upset, asked how I could be ignorant of the most famous laser spectroscopist in the world? My ignorance turned out not to be true though, although I was confused at the moment, as Ahmed and I found out when we met in my office some weeks later. I pulled the Swedish National Encyclopedia from the shelf, and there was a several-pages-long article about Physical Chemistry, authored by myself some five years earlier, featuring molecular reaction dynamics studied by fast laser spectroscopy. It contained an illustration from one of Ahmed's papers! Furthermore, according to Ahmed and Ingmar, I should have said that he (Zewail) would be warmly welcome, provided costs for his travel to and stay in Gothenburg were taken care of by Grenthe or the Academy. I translated my encyclopedia article for Ahmed, which initiated our intense discussion about where Physical Chemistry was heading. I would say we found each other then — I even paid his tram ticket when we, at sunset, went to my home where my surprised wife, Gunnela, improvised dinner for us. Ahmed stayed long that night, and many basic questions that we touched upon then and there, we have returned to at our encounters later at Caltech.

After his Nobel prize, we got closer, partly because of our joint interest in molecular spectroscopy but partly also, I think, because of a kind of isolation Nobel laureates often feel and me being a member (at the time Chair) of the Nobel Committee. Both of us felt a pressure to behave and not openly admit any of those embarrassing questions and doubts that incessantly plague scientists who want to reach for true understanding of how things really work. We both felt that the textbook word and reality are more often than not two different things — a little like a priest who might have his private doubts. Both of us believed in intuition as the best guidance to new theories and experiments and also allowed ourselves to be secretly skeptical to many "accepted facts" in science.

I used to visit Ahmed at Caltech for a couple of days each November, and one spring I stayed for several months invited as a visiting scholar in his lab. I am very grateful to him (and his wonderful wife Dema) for their ever-generous and amiable hospitality during these stays, which were truly inspiring to me. The schedule of the first day of my visits was always that we sat down to discuss various things that had happened in our respective research lives since last time — mostly dominated by Ahmed's often much more spectacular experimental progress and discoveries. We always started our day with a cappuccino together under the sun outside the campus cafeteria. There he would ask about things related to the Nobel prize, where he, although he was a Foreign Member of the Royal Swedish Academy of Sciences, was often ignorant about who were nominated and so on. (Foreign Members are welcome to those meetings in the Class where nominations and investigations of various candidates are presented and discussed, but have not the right to participate and vote at the general assembly finally deciding on the Nobel prizes and are normally not informed about any details of ongoing investigations, who are hot candidates, etc.). Ahmed cared very much for the status of the Nobel prize and sometimes had his concerns about true or suspected candidates. This "updating" talk I found the most difficult in our relation as Ahmed's natural (scientific!) curiosity felt no bounds, while my own integrity and responsibility toward the Nobel institutions prevented me from a completely open information transfer. At the same time, these discussions were very useful for me as Ahmed was always an updated source of current developments, indications of upcoming breakthroughs, paradigm shifts, and so on. Our own fields were thoroughly ventilated, but any development important to Chemistry as a whole was also considered.

Moving from the coffee table to Ahmed's office, or rather his big conference table and whiteboard across the hallway which we filled with sketches and equations, we plunged into science. Two topics were recurrent: diffraction of singular electrons in a Young's double-slit setting and coherence of molecular vibrations. I was very much a skeptical Dr Watson, when Ahmed presented new results proudly like a cat putting a dead mouse at my feet. He also enjoyed using me as a Devil's Advocate: for example, I was generally suspicious about visible oscillations, after having revealed some artifacts in laser spectroscopy back in the 1980s, one being a ringing effect due to thermal-lens fluctuations instead of a real claimed molecular orientation effect due to interaction between the photon field and induced molecular dipole moments, the so-called Optical Kerr Effect. Very early his elegant demonstration of recurrent wave-packet travels in the potential diagram of sodium iodide was a focus of our discussions. Was his result probably in conflict with Heisenberg's uncertainty principle? The answer we finally agreed on was, however, "no," because the fact that Ahmed was exciting all the NaI molecules in phase would make the sample behave differently from a quantum mechanically described single molecule (microscopic system), and be more like a coherently vibrating macroscopic system.

A pedagogical device, constructed of Perspex and bent steel strings by my engineer Mr Tore Eriksson, and which Ahmed used in his Nobel Lecture illuminated on an overhead projector, showed how a steel ball was moving back and forth in the excited-state potential of Na–I. When the potential curve came close to the ground-state potential, the ball could jump over to the ground state, triggered by a hand-controlled mechanical switch, and then either return to the deep valley representing the bound state of the molecule or move in the opposite direction, corresponding to dissociation into Na and I atoms. Ahmed was very fond of the device, which had an honorary position on a shelf in his office.

Ahmed's other great interest, in single-electron diffraction, of course, was the embryo to his next big discovery: the use of time-resolved electron microscopy to resolve fast processes. We devoted a lot of time to discuss whether the Coulombic repulsion between the electrons would create a longitudinal "anti-bunching" along the train of electrons, or would just spray the electrons randomly in space.

As Ahmed's guest at Caltech, I had the privilege and pleasure of meeting his friends at the "Round Table" Wednesday lunches at the Athenaeum, 10 of Caltech's most outstanding scientists, including Rudolph A. Marcus and Jack Roberts, the latter who passed away in October 2016, at 98, was the legendary inventor of physical organic chemistry and the modern use of NMR for studying molecular structure and dynamics. To a newcomer, these lunches were indeed inspiring and also a little scary — I suddenly had to stand up for all science-, politics-, and economy-related affairs that had been going on in Europe since my last visit; generally I managed to duck for the kill questions, although not always. At these Round Table discussions about all and everything, I also realized how Ahmed was the born survivor — a fish in his right element — the same way as he survived when he was first up for interview at Caltech and became afraid he would misspell Feynman when writing his name on the blackboard; he got away with it by, after writing Fe..., turning around with a charming smile saying: "we all know how to spell Feynman, do we not?" whereupon everyone laughed.

To Ahmed Zewail, the coming generation of young scientists was very important — and Egypt's youth had a special place in his heart — as will Egypt be forever grateful to Ahmed Zewail for all his influence and academic initiatives there, such as the Zewail City of Science and Technology, his lifelong dream, inaugurated with Ahmed as its Chair in 2011. He also played a seminal role for the early development and establishment of the Molecular Frontiers Foundation (MFF), being Chair of its first Scientific Advisory Board and a speaker at many of its symposia, explaining results from the research frontiers in his characteristic vivid, simplistic, and pedagogical way. MFF owes Ahmed a lot in that he created the discussion atmosphere that today is a hallmark of the organization: the youth are prized for their questions, not for knowing the answers!

My first experience of this talent of his was at a Lindau Nobel Laureate Meeting where I was asked to moderate a round-table discussion, an occasion that gave rise to another humorous anecdote that Ahmed liked to refer to. The Nobel laureates around the table, in addition to Ahmed, were Harry Kroto, Paul Crutzen, Richard Ernst, and George Olah (see Figure 7.1). I received scraps of paper with questions from the floor — mostly young Ph.D. students who had received stipends to partake in the meeting. As most questions were thoughtful and scientifically advanced (read boring), I smuggled myself in a few that fired up the discussion considerably: referring to Alfred Nobel, one asked how dynamite really works? What is the role of the stabilizing agent (a dispersion of dry diatom silica organisms or simply sawdust)? A lively discussion started among the Nobel laureates with speculations all over the place, one was about acoustic damping, until it was interrupted by a sudden shout from the audience — Nobel laureate Manfred Eigen rose and claimed we were all wrong, and the true explanation how the tricky nitroglycerine was tamed was that the stabilizer reacts with radicals, quenching chain reactions.

We were interviewed afterward by the Editor of *Chemical & Engineering News*, Dr. Madeleine Jacobs, who, when she introduced herself, turned to me and said: "we have actually met before but maybe you don't recognize me with clothes on." This comment, which led to

Figure 7.1. Lindau Nobel Laureate Meeting 2002. From left to right: Harry Kroto, Ahmed Zewail, George Olah, myself, Richard Ernst, and Paul Crutzen. Reproduced with kind permission from Madeleine Jacobs and the C&EN.

great general amusement, not least Ahmed's, had an explanation in that both of us had started our days with a swim workout in the hotel pool before breakfast. After a couple of days, we greeted each other at a distance, being the only ones in the large pool but I had obviously not recognized her when she reappeared in the conference without swim cap and "with clothes on." Another amusing incidence occurred when I was going to check out from the hotel and had no cash (they refused to take any cards). Behind me in the line was Dr. Lorie Karnath, restless as she had to catch a flight, and possibly also feeling pity for me: she bailed me out and paid my room and thus probably rescued me from an embarrassing fate. In return, I later treated her to a nice meal and great wine at the Grand Hotel in Stockholm. Together with Prof. Magdalena Eriksson, she later became instrumental during the first stumbling, founding steps of creating MFF; both of them are still highly active in the core of the MFF organization.

A special memory I have with Ahmed was one summer afternoon when he was on his way from Lund to a spectroscopy conference in Copenhagen, and made a detour via our summerhouse at a little fishing village by the Sound between Sweden and Denmark. He stayed for an early dinner but as he was going to call for a taxi to take him to the hydrofoil ferry in Malmö, I suggested: why don't we sail you over and you save some money? Soon we were under sail with my neighbor and very close friend, orthopedic professor Björn Persson, in his boat with course straight west. The wind was not cold, but brisk and we sailed fast. After a little more than two hours we entered the mouth of the harbor of Copenhagen, reduced sail and moved slowly into the very heart of the continental city. Both Ahmed and I remembered the event fondly, this was the first time for him in a sailing yacht — his only experience of small boats being from the Nile when he was a boy. He enjoyed every moment, especially the magic feeling when we were silently approaching the big city in darkness with its many lights ashore. We landed in Christianshavn, a part of Copenhagen which is a little like Venice with narrow channels and many small restaurants. It was close to midnight and we wondered if it would be possible to get anything to eat before Ahmed had to depart for his hotel and we return to Sweden. Close to where we had docked the boat, we found a small pub in the basement which was closing with the last patrons just leaving. When Ahmed asked the owner whether we could have something very simple to eat, he was most reluctant, but Ahmed's charm quickly broke down any resistance, and soon we found ourselves sitting together with him and the kitchen staff having a veritable feast meal — one of the waiters (a Portuguese) brought his guitar and began to entertain us all with emotional Fado songs.

The last time I saw Ahmed was when Gunnela and I visited Pasadena in November 2015. I gave a seminar in his group on our recent discovery of a new elongated conformation of double-stranded DNA (the "sigma form") and its potential biological role and why any form of organic life can only exist in a water-rich environment. He had many good points and also suggested a clever mechanistic complement to my explanation of the inhomogeneous conformational reorganization that we call "disproportionation." Earlier the same day, he had shown me some amazing results from time-resolved single-electron microscopy, visualizing in real time

Figure 7.2. Ahmed and I at the Molecular Frontiers Symposium in May 2015. Reproduced with kind permission of the photographer, Mr Jonas Ekberg.

the coherent travel of phonons along a set of hydrocarbon chains aligned at a surface. At the end of the chain, the wave was reflected and returned back. He commented when I interrupted him to guess that result: "you saw it in a femtosecond, didn't you?"

The last photo I have of Ahmed (and myself) is from my 70th birthday symposium in Gothenburg in early May 2015 (Figure 7.2). As usual, he gave a captivating talk (to be found live on www.MolecularFrontiers.org), appreciated as much by the professional scientific audience (many Nobel Laureates) as by the more than 200 high-school students.

Ahmed, I am most grateful to you for all inspiration you gave me over the years and for your deep-felt friendship, the memory of which I will carry in my heart as long as I live.

Author Biography

Bengt Nordén is a Swedish chemist and molecular spectroscopist. He is Chair Professor of Physical Chemistry at Chalmers University of

Technology in Gothenburg, His contributions cover a range of fields in science: from fundamental properties of aromatic chromophores and methods for determining transition moments, to topics related to biomolecular recognition and 3D structure of DNA complexes assessed with polarized light methods. His studies inspired the design of several, now famous constructs, including peptide nucleic acids and DNA threading intercalators with unique kinetic selectivity. Nordén is promoting science globally to a general public through Molecular Frontiers (www.molecularfrontiers.org), and as Chairman of The Nobel Committee for Chemistry, he developed tools for identification of breakthroughs in science with impact also in fields outside chemistry. Due to his scientific achievements, he is the recipient of some of the most distinguished awards and biggest individual grants in Europe. Nordén is a member of ten Academies in Europe and an Honorary Fellow of the Royal Society of Chemistry (HonFRSC), an Honorary Member of the Chemical Societies in China and India, an Honorary Fellow of the Australian National University, and a foreign member of the World Academy of Sciences for the advancement of science in the developing countries (TWAS). He was awarded recently an honorary degree by the President of Singapore for contributing to moving one of their universities to the absolute world top.

Chapter 8

A Tribute to Ahmed H. Zewail

*David C. Clary**

In the 1990s, I was fortunate enough to begin a scientific interaction with Ahmed Zewail that led to a deep personal friendship. His research group had performed time-resolved experiments on the chemical reaction between OH and CO, which forms a short-lived "HOCO" complex.[1] We had performed quantum scattering calculations on this reaction, which gave microsecond lifetimes for HOCO, agreeing quite well with Ahmed's experimental measurements.[2] As an Editor of *Chemical Physics Letters*, Ahmed strongly believed in close interaction between theory and experiment and, initially through this research, I often found myself speaking in the same sessions with him at conferences. This included the Nobel Symposium on Femtochemistry in 1996, where we both had the privilege of giving a lecture at Alfred Nobel's own library in Björkborn, Sweden.[3] Following this, Ahmed arranged for me to speak at the international series of conferences on femtochemistry, which were started by Jörn Manz in 1993 and were very close to Ahmed's heart.

It did not escape the notice of those of us speaking at the Nobel Symposium that several of the members of the committee for the Nobel Prize in Chemistry were attending, and there was a strong feeling that a Nobel Prize for Ahmed must be awarded very soon. So when the

---

*david.clary@magd.ox.ac.uk

announcement finally came from Sweden three years later, there was much rejoicing in the femtochemistry community.

Ahmed was very much an anglophile and was a regular visitor to the UK, where I met him many times in Cambridge, Oxford, and London. My wife Heather and I remember meeting him with Dema and the boys during a nice sabbatical summer they spent in Cambridge. There Ahmed gave a series of research lectures in the Cavendish Laboratory of Physics. This is the laboratory of Maxwell, J. J. Thompson, Rutherford, Bragg, and Crick and can be a highly critical audience. One of the top professors there told me that the first research lecture in a series at the Cavendish Laboratory is quite often excellent but the subsequent ones are often not nearly so good. However, he found that each of Ahmed's lectures just got better and better. Ahmed was pleased to hear this observation — it was deep scientific praise.

Amyand David Buckingham, Professor of Theoretical Chemistry at Cambridge, was the Editor of *Chemical Physics Letters* along with Ahmed for many years. David retired from that role in 2000, and I was fortunate enough to be invited to replace him as editor. Consequently, I met Ahmed regularly at meetings to discuss the journal. These meetings usually occurred in Amsterdam in September, where we were taken to interesting restaurants by the publishers Elsevier. Ahmed's international status and fame, especially among people from his home country of Egypt, was becoming apparent and I recall him often being stopped in the street for his autograph or a photograph just like a rock star. Ahmed usually took these trips to Amsterdam as an opportunity to fly back to Egypt after the editorial meetings to visit his family. I recall being with him on the morning of September 11, 2001, at Schiphol airport where he was taking a plane to Egypt and I was returning to the UK. Little did we realize the tragic events that were to occur on that day in the USA.

Ahmed was elected a Foreign Member of the Royal Society in 2001. There are very few such elections and he was honored to add his signature to the Charter Book signed previously by such scientific greats as Isaac Newton and Charles Darwin. Not long after, I was involved in jointly organizing a Royal Society Discussion Meeting on Energy Landscapes. Ahmed gave the opening talk, and it provided the opportunity for him to speak on the new research area he was developing: "Diffraction, crystallography and microscopy beyond three dimensions: Structural dynamics in space and time".[4] Many of us feel this pioneering work was worthy of a second Nobel Prize.

I moved to Oxford University in 2002 and was delighted to nominate Ahmed for an Honorary Degree which he was awarded in 2004. The ceremony to bestow this honor was held at Christopher Wren's historic Sheldonian Theatre. It involved a long oration in Latin on Ahmed's achievements[5] and finished with the statement from the Chancellor of Oxford University, Lord Patten:

> "Molecularum metitor accuratissime, atomorum existimator ingeniosissime, cuius egregiis laboribus tam alte absconditas veritatis formas cognoscere discimus, ego auctoritate mea et totius Universitatis admitto te ad gradum Doctoris in Scientia honoris causa."

which in translation states:

> "You are most accurate in the measurement of molecules, most inventive in your observation of atoms. Your outstanding work has enabled us to understand these profoundly hidden areas of truth. Acting on my own authority and on that of the University as a whole, I admit you to the honorary degree of Doctor of Science."

Following this, Ahmed was keen for the Femtochemistry Conference to come to Oxford, and I was delighted to host the meeting in 2007 at Magdalen College, where I had become the President.[6] At the conference dinner, we were pleased to present Ahmed with a fine drawing by the artist Peter Edwards who had previously drawn or painted the Nobel Prizewinners Erwin Schrödinger and Seamus Heaney, both Fellows of Magdalen College (see Figure 8.1).

In the library at Magdalen College, we have a collection of first editions of many great scientific books, and Ahmed was very interested to examine our original copy of Robert Hooke's remarkable *Micrographia*. This was the first book published by the Royal Society in 1665 and essentially introduced the world to the use of the microscope in detailed scientific study. Ahmed then often used the iconic pictures from this book in

Figure 8.1. Drawing of Ahmed Zewail by Peter Edwards (2007).

his own scientific talks on microscopy and published "*Micrographia* of the twenty-first century: From *camera obscura* to 4D microscopy," as part of the 350th Anniversary celebrations of the Royal Society.[7] Ahmed invariably gave an historical feel to his lectures and speeches, and he was very proud to include examples of the great heritage of Ancient Egypt. Figure 8.2 records Ahmed's visit in front of the medieval tapestries in my President's Lodgings at Magdalen College.

Ahmed always gave top priority to science, but around this time, he was becoming concerned with developments in the Middle East. While in Oxford, he gave a lecture to the Centre for Islamic Studies, where he stated that if a fraction of the money used for military activities in the Middle East is spent instead on education, then things would improve dramatically in that region. These thoughts may have been on his mind when he accepted the invitation of President Obama to be the first US Science Envoy to the Middle East. In the same year, I was made the first Chief Scientific Advisor to the UK Foreign and Commonwealth Office, and our paths crossed in these ambassadorial roles. We both spoke at the World Science Forum in Budapest, where the theme was on how scientific interactions can enhance

Figure 8.2. David Clary and Ahmed Zewail at Magdalen College Oxford in July 2007.

international relations, an activity known as Scientific Diplomacy.[8] It was the first time I had heard Ahmed give a talk at a conference in which he did not mention femtochemistry, and it was also the first time he had heard me speak without referring to quantum mechanics!

Ahmed retired as Editor of *Chem. Phys. Lett.* in 2007, and the Editors were delighted to establish the Ahmed Zewail Prize in Molecular Sciences with the publishers Elsevier. The prizes have been often presented biennially at meetings of the American Chemical Society, and Ahmed frequently spoke at the symposia associated with the awards. The Prize has been presented so far to Amyand David Buckingham, Mostafa El-Sayed, Bill Miller, John Meurig Thomas, and Noel Hush, who are all pioneers in different areas of chemical physics. In August 2017, the prize will be presented to Michael Graetzel at the 13th Femtochemistry Conference in Cancun, Mexico, which will be dedicated to the legacy of Ahmed Zewail.

My last communication with Ahmed was just one month before he died. I had been fortunate enough to be knighted by the Queen, and Ahmed sent me his very kind congratulations. Even though he clearly was not well at this time, he still made what must have been a significant effort to send his message. Ahmed will be remembered not only as a great scientist but also as a most remarkable person.

References

1. N.F. Scherer, C. Sipes, R.B. Bernstein and A.H. Zewail, "Real-time Clocking of Bimolecular Reactions: Application to H+ CO_2," *J. Chem. Phys.* **92**, 5239 (1990).
2. D.C. Clary and G.C. Schatz, "Quantum and Quasiclassical Calculations on the OH + CO → CO_2 + H Reaction," *J. Chem. Phys.* **99**, 4578 (1993).
3. V. Sundström (Ed.), *Femtochemistry & Femtobiology: Ultrafast Reaction Dynamics at Atomic-Scale Resolution*, Nobel Symposium Monograph (World Scientific, Singapore, 1996).
4. S.M. Pal, J. Peon and A.H. Zewail, "Biological Water at the Protein Surface: Dynamical Solvation Probed Directly with Femtosecond Resolution," *Proc. Natl. Acad. Sci.* **99**, 1763 (2002).
5. See: http://www.ox.ac.uk/gazette/2003-4/supps/1_4702.htm#7Ref.
6. D.C. Clary (Ed.), "Femtochemistry and Femtobiology," *Chem. Phys.* **350**, 1 (2008).
7. A.H. Zewail, "Micrographia of the Twenty-First Century: From Camera Obscura to 4D Microscopy," *Phil. Trans. Roy. Soc. A* **368**, 1191 (2010).
8. See: http://mta.videotorium.hu/en/recordings/1463/science-diplomacy-at-the-uk-foreign-and-commonwealth-office; http://mta.videotorium.hu/en/recordings/1486/speech-of-ahmed-zewail.

Author Biography

Sir David Clary has been President of Magdalen College, Oxford, since 2005. He also directs a research group at the Department of Chemistry, University of Oxford, working on the quantum theory of chemical reactions. He has been Editor of *Chemical Physics Letters* since 2000, a role he shared with Ahmed Zewail for several years. Previous appointments include being Head of the Division of Mathematical and Physical Sciences at Oxford University, Director of the Centre for Theoretical and Computational Chemistry at University College London, and Reader in Theoretical Chemistry at the University of Cambridge. He received his Ph.D. from Cambridge and his B.Sc. from the University of Sussex. He has been elected to several academies, including the Royal Society, the American Academy of Arts and Sciences, the American Association for the Advancement of Science, and the International Academy of Quantum Molecular Science. He was President of the Faraday Division of the Royal Society of Chemistry from 2006 to 2009, and was the first Chief Scientific Adviser to the UK Foreign Office from 2009 to 2013. In 2016, he was knighted by Queen Elizabeth II for his services to international science.

Chapter 9

Ahmed Zewail — A Great Scientist and Inspiring Friend

*Amyand David Buckingham**

It is a special honor to be invited to contribute to this volume honoring a great scientist, a colleague, and a very good friend. My first acquaintance with the name Zewail came through a 1971 paper in *Chem. Phys. Lett.*[1] It was written with his Ph.D. supervisor at the University of Pennsylvania, Robin M. Hochstrasser. Robin, a graduate of the University of Edinburgh, spent his sabbatical leave in 1972 with our Theoretical Chemistry Group in Cambridge. Both he and I were original members of the Advisory Editorial Board of *Chem. Phys. Lett.* when the journal started in 1967. He went on to become an Editor of *Chem. Phys.* In 1971, I was preparing a review article on the Stark Effect and wrote: "Hochstrasser and Zewail observed a first-order Stark splitting of the degenerate $^1E''(\pi^* \leftarrow n)$ states of *s*-triazine, s-$C_3H_3N_3$, in a single crystal at 4.2 K when the field is in the plane of the molecule. The effect is of special interest since it arises from a similar source to that in the hydrogen atom, that is, from a field-induced mixing of degenerate orbitals; it is the first example of a first-order Stark effect in a non-polar aromatic molecule".[2] I was very impressed with that paper as well as with the style of the research.

---

*adb1000@cam.ac.uk

Ahmed's research style — its conception, execution, and presentation — was outstanding and truly exemplary.

It was through the journal *Chem. Phys. Lett.* that I became well-acquainted with Ahmed. I was appointed Editor in 1978 following the death that year of one of the two Founding Editors, Jan Hoytink. I joined the other Founding Editor, Laurens Jansen of the Battelle Institute in Geneva, and Richard B. Bernstein, then at Columbia University in New York, who was the first American Editor. Laurens Jansen was suffering from health and other problems; he died in 2005. Ahmed joined the Advisory Editorial Board in 1985 and became Acting Editor in 1990, and Editor in 1991, following the sudden death of Dick Bernstein while attending a USA + USSR Laser Conference in Russia in July 1990. Dick had been working in Ahmed's femtochemistry laboratory at CalTech — he saw the beauty, power, and potential of that research and wanted to play a part in it (see Figures 9.1 to 9.4). Ahmed's noble offer to step into the breach and maintain the American editorial office of *Chem. Phys. Lett.* was typical of the man. He and I worked together very successfully as Editors for ten years. We corresponded frequently and met at the annual meeting with the Publishers in Amsterdam. We also met at the annual meeting of the

Figure 9.1. A photograph taken in July 1990 in Ahmed's office at CalTech.

Ahmed Zewail — A Great Scientist and Inspiring Friend 57

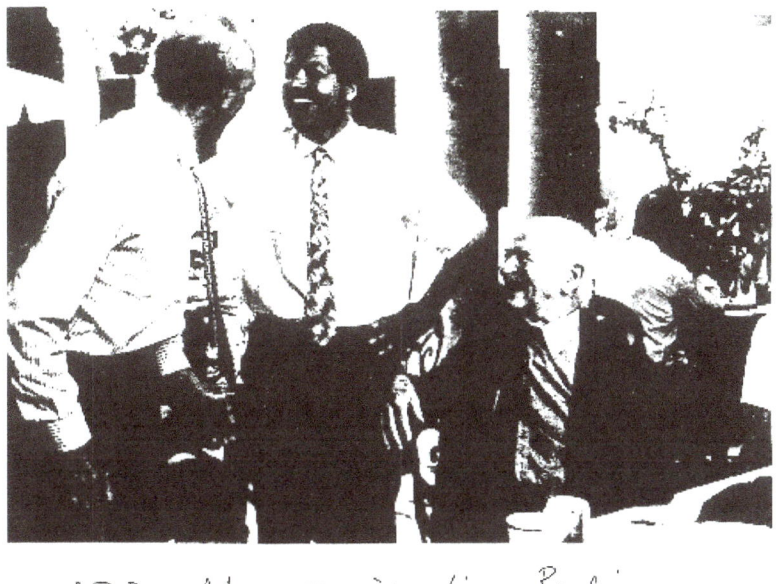

ADB, Ahmed Zewail, Linus Pauling

Figure 9.2. Ahmed and ADB talking to Linus Pauling in Ahmed's office at CalTech in 1990. Ahmed was the first Linus Pauling Professor of Chemical Physics at the California Institute of Technology. We were discussing the dissociation energy of alkali-halide diatomic molecules, such as NaCl.

Figure 9.3. Ahmed with Jill Buckingham at a dinner in Amsterdam in 1994.

58 *A. D. Buckingham*

Figure 9.4. Ahmed with Dema Zewail, ADB, and Patrick Jackson of Elsevier at a dinner in Amsterdam in 1998.

National Academy of Sciences in Washington, at conferences and meetings, and on occasions in Cambridge and Pasadena. I had huge respect for him — for his intellect, his judgment, his warmth, and his wisdom. Ahmed was a true life-enhancer.

He gave a brilliant talk at my retirement symposium in Cambridge in 1997 and another at the 233rd ACS National Meeting in Chicago on March 26, 2007, when I was lucky enough to receive the first Ahmed Zewail Prize in Molecular Sciences.

More than once Ahmed visited our large house in Snailwell near Cambridge. He called it "The Buckingham Palace"! My wife and I miss him and Dema greatly. His name, his family, and his science will continue to live on.

References

1. R.M. Hochstrasser and A.H. Zewail, "Stark and Zeeman Effects on the Singlet nπ* State of S-Triazine," *Chem. Phys. Lett.* **11**, 157–158 (1971).
2. A.D. Buckingham, "The Stark Effect," in *Physical Chemistry*, Vol. 3, *MTP International Review of Science*, D.A. Ramsay (Ed.) (Butterworths, London, 1972), pp. 73–117.

Author Biography

Amyand David Buckingham was born in Sydney, Australia, in January 1930. He graduated from Sydney University with a B.Sc. degree and University Medal in 1951 and earned his M.Sc. degree in 1953, having been supervised by Raymond Le Fèvre. He was a Shell Postgraduate Scholar at Cambridge University from 1953 to 1955, gaining a Ph.D. degree under the supervision of John Pople (he was Pople's first research student). In 1955, at the Physical Chemistry Laboratory at Oxford University, he held a Senior Research Studentship of the Royal Commission for the Exhibition of 1851. He was Lecturer in the Inorganic Chemistry Laboratory at Oxford from 1958 to 1965, Professor of Theoretical Chemistry at Bristol University from 1965 to 1969, and Professor of Chemistry at Cambridge University from 1969 to 1997. He is an Honorary Fellow of Pembroke College, Cambridge; a Fellow of the Royal Society; a Corresponding Member of the Australian Academy of Science; a Foreign Associate of the National Academy of Sciences; a Foreign Honorary Member of the American Academy of Arts & Sciences; a Foreign Member of the Royal Swedish Academy of Sciences; and a Member of the

International Academy of Quantum Molecular Science. He received the Faraday Medal of the Royal Society of Chemistry in 1998, the inaugural Ahmed Zewail Prize for Molecular Sciences in 2007; and the C. H. Townes Award for quantum optics from the Optical Society of America in 2001. He was Editor of *Molecular Physics* from 1968 to 1972 and of *Chemical Physics Letters* from 1979 to 1999. He has published over 340 papers in scientific journals.

Chapter 10

The Pyramid Builder

Majed Chergui*

"Did you bring *Al Antabli*?" Ahmed would ask me cheerfully every time we met and settled for a chat. Despite his undeniable pedagogical skills, I never got from Ahmed what *Al Antabli* (العنتبلي) really meant. Only recently did I learn that it was a word in Alexandrian dialect that could mean anything as long as it was positive and pleasant. In the context of my encounters with Ahmed, I knew he meant Cuban cigars. These were to be part of all our meetings.

I first met Ahmed in 1988 at the Molecular Electronic Spectroscopy conference in Ile d'Oleron (France) and was struck not only by his brilliant talk and dynamism but also by his contagious optimism and uplifting spirit. I was then a postdoc in Berlin, and beyond the fact that this encounter was to completely change the course of my future career, it was also the start of a deep friendship, not to say the least, a brotherly relationship. From then on, we regularly met two or three times a year whenever Ahmed visited Europe or I visited the US. Our meetings were quickly marked by a ritual of long and deep discussions about science, politics (Middle Eastern in particular), and life in general. These were moments of

---

* majed.chergui@epfl.ch

total openness, where all subjects were discussed frankly and in a very friendly way (with the unavoidable *Antabli*). Ahmed had a knack of making people feel at ease and open their heart, irrespective of their background. But I also felt that because of our common Middle Eastern origin, we shared the same concerns and hopes, and the same culture, which included a certain sense of humor.

For example, after 2000, every time we met we would greet each other by: "Doctor, I have back pain!" The story goes that following Ahmed's Nobel Prize win for Chemistry in 1999, a symposium was organized in his honor by Marcos Dantus at the 2000 ACS Spring meeting in Washington. I was invited to chair the opening session. Ahmed, Dongping, and I met the night before and went for dinner to a Lebanese restaurant that Ahmed knew in Georgetown. The place was nothing posh, but the food was excellent and Ahmed was in high spirits, asking jokingly to the waiter what good "*Antabli*" he would recommend for dinner? The waiter, a Lebanese man well in his 60s, was evidently amused by Ahmed's cheerful mood and his Arabic Egyptian accent. At some point, a group of customers next to our table stood up to leave and one of them recognized the famous Ahmed Zewail. They all started to greet him (in Arabic) and shake hands with him: "Doctor, you are the pride of the Arabs…," "Doctor, congratulations, it's an honor to meet you…," and so on. When the group left, the waiter came to our table and said to Ahmed: "so you're a doctor," Ahmed replied in the affirmative, and the waiter went on, "Oh! You know doctor, I have terrible back pain, do you have a cure for it?" Ahmed's generous laugh resounded in the restaurant, as I explained to the waiter that he was not that type of "doctor."

When I think of Ahmed, in addition to the genius who revolutionized Science, I remember his gentleness, generosity, truthfulness, respect for the others, magnificence, charisma, and amazing sense of humor. The above anecdote is just one of the many examples. Most of all, there was no separation between the Man and the Scholar, between his passion for Science, his *Weltanschauung*, and his *joie de vivre*.

At a gathering of a group of participants at the International Conference on Photochemistry in July 1991 in Paris, Ahmed said: "I will ask each of you what his deepest wish in life is?" When his turn came, he described to us how he wanted to leave a mark in Science and to create a momentum for science in Arab societies. Thinking in retrospect of this conversation, it strikes how Ahmed could see through his Life to come. It is as if he knew exactly what he would be doing, scientifically and otherwise, in 10 or 20 years from that moment. He was already a celebrity in Science following his breathtaking achievements during the past four years at that time. A revolution in Science was unfolding, which he had triggered. It would culminate in the first Femtochemistry conference that was organized by Jörn Manz in Berlin in March 1993 (Jörn gives a detailed account of this event in his beautiful contribution to the present book).[1] This was a special event, not only due to the participation of world class scientists (Rudolph A. Marcus and John Polanyi were present) but also because of the excitement of the new science. It also consecrated the field, whose name was coined by Ahmed himself.

Later that year, I moved to Lausanne to take up a Faculty position in Physics and, inspired by Ahmed, start my own research group in Ultrafast Science (that was not my field of research before). I invited Ahmed to give a Public Lecture to the Société Vaudoise des Sciences Naturelles in Spring 1994. His lecture was admirable, it galvanized the audience and created a lot of excitement. It was on the way back to my office after his Lecture that he suggested I organize the next Femtochemistry conference. I was barely starting and was a no-name in the field; I objected, but he quite rightly argued that this would be an opportunity to get into it. I can never be grateful enough to him for the trust he put on me, as it really set me on track.

The Lausanne conference took place in September 1995.[2] For a field that was barely eight years old, the attendance was high (about 250 participants) and the giants of the field were present (Ahmed, Vladilen Letokhov, Kent Wilson, J.T. "Casey" Hynes, Shaul Mukamel, Douwe Wiersma, Steve Berry, Ken Eisenthal, Joshua Jortner, Jörn Manz, Will Castleman, and many others), while many of the young attendees who were either Ph.D. students or postdocs at that time have now become leaders in the field. There was magic in the air, which needless to say, Ahmed fueled by maintaining the high intellectual level of the debates sprinkled with good humor (see Figures 10.1 and 21.1). For him, the priority was on exchange and debating of ideas, regardless of the rank or status of the discussion partners.

Figure 10.1. The author and Ahmed Zewail at the Banquet of the Femtochemistry Lausanne Conference in 1995. To the left is the author's wife, Marie-Agnès.

Ahmed was to become a regular visitor to the University of Lausanne through the 1990s, involving himself deeply in discussions with my colleagues and opening the "temporal landscape" to chemists, physicists, and biologists. I nominated him for the Honorary Doctorate of the University, arguing that it would be important for the University to award it to him before he gets the Nobel Prize, a conviction I and many of my colleagues shared firmly. He received the Honorary Degree of the University of Lausanne in 1998. A year later, he was awarded the Nobel Prize for Chemistry.

The awareness of nuclear dynamics in describing phenomena in chemistry, biology, and materials science, that was brought about by the Femtochemistry "revolution," was to trigger the "dawn of a new Era," as John Thomas wrote in a comment[3] to an article entitled "Structural Femtochemistry" published by Ahmed in 1991,[4] in which the latter laid the foundation of Structural Dynamics with ultrashort pulses of electrons. An alternative approach that was also considered at that time was to use X-rays. The latter was adopted by most of the community (myself included) but Ahmed was the sole person to choose electrons and was the first to deliver snapshots of molecular reactions.[5] In his Nobel lecture, Ahmed thanked, not devoid of a touch of humor, the Nobel Committee for not mentioning his work on ultrafast electron diffraction in its citation. Here again, he knew well ahead of time what he was stepping into, and the spectacular results did not take long to appear, triggering in the following years another revolution with groundbreaking discoveries. This opened a new landscape in Science, especially after the advent of what he named four-dimensional ultrafast electron microscopy (4DUEM).[6] I have no doubt that he would have received a second Nobel Prize for his work on time-resolved electron methods.

Despite our long relationship and frequent scientific discussions, we only published one review paper together, in 2009.[7] In my opinion, this was the first paper discussing the achievements of both ultrafast electron and X-ray methods for chemical systems, and I have to confess that electron-based methods were ahead of the X-ray ones. While writing it up, I was impressed by Ahmed's high standards for exactitude on every point. He had an amazing ability to focus on a problem and finish the tasks to accomplish with no time ("in a femtosecond!" he would say). I had previously experienced this amazing ability of his to concentrate on a problem, without letting himself be distracted, a couple of days after he had received the Nobel Prize. As soon as the news came, I first sent him a fax of congratulations, then I repeatedly tried to reach him by phone but was unsuccessful. I decided to wait for another 24 hours and tried the next day. This time I did manage to talk to his secretary, who told me he could not attend me as he was going through an article with two of his postdocs! I was flabbergasted! But I was also moved when he called me the following day at 10 AM (it was 1 AM in California) to apologize for not attending my phone call and to thank me for the fax and the call. Why would he bother to call back when he literally "had the entire world trying calling him," as he quite rightly said?

Ahmed emerged as a Man of Wisdom in the sense of the first University in History, "Beyt Al Hikmah" or House of Wisdom, which was founded in Bagdad in the seventh century by the Calife Al Ma'mūn. Like scholars of the medieval times and the Renaissance, he was a man of vast culture with an amazing imagination, creativity, vision, and no boundaries to his curiosity. We often talked about history and, in particular, history of science and he was curious to trace back the origin of scientific quest in the Arab civilization over a thousand years ago. He often referred to the tenth century scholar, Ibn-Al-Haytham, who initiated the experimental method as we practice it today. His approach of a given scientific question was never disconnected from its potential implications at a practical level. A good example is given by the title of the beautiful book he edited in 2008: *From Atoms to Medicine*.[8] The book bears the mark of his perfectionism and is an admirable example of his ability to embrace and blend together problems that seem difficult or almost impossible to link together.[9] But his insatiable curiosity and his eagerness to learn and understand went beyond Science. He read a lot and was always intellectually alert to learn something new. His wisdom also made him ponder and weigh carefully how to judge a situation, an event, or a person. It also made him seek what brings people together rather than what divides them. He was quite touched by the film The Destiny, by Egyptian film maker Youssef Shahine that described the fate of medieval scholars in the Muslim and Christian worlds, trying to build ties and a common culture of Knowledge and Thought beyond political and religious restrictions. Later, the same Youssef Shahine was to highlight Ahmed in another film. He also wanted to understand societies and how they change and evolve, maybe in reference to his own, to which he was deeply attached.

Ahmed grew up in the Egypt of the 1950s and 1960s. At that time, Egypt represented the lighthouse of the Arab world in the cultural and political arenas. Its cinema, music, and literature largely dominated Arabic culture in quality and quantity. Cairo was a center of intellectual life and in Science, Egypt had the best Universities. Politically, President Nasser was trying to restore the Pride of the Arabs after decades of colonialism, and he enjoyed immense popularity in the Arab world.

Ahmed often spoke about the humbleness and genuineness of human relations in the secure and stable social and family environment he had grown in. This forms the basis of his deep roots in Egyptian culture, which he cherished from the depth of his heart. It is often said that the highest trees have the deepest roots. His rise to excellence and magnificence was, to my eyes, a trait of his deep roots into Egyptian culture. Indeed, preceding him or contemporary to him had been personalities such the physicist Sameera Moussa, writers such as Taha Hussein or Naguib Mahfouz, or film producers such as Youssef Shahine, mentioned earlier.

But most important was Oum Kalthoum, the diva of Arabic music, also called the "Star of the East" or the "Fourth Pyramid," who died in 1975. She enjoyed, and still does, an immense popularity in the entire Arab world and beyond. Ahmed had a great admiration for her and was transported when listening to her music. I still recollect an evening at home after the Lausanne Femtochemistry conference, which we spent listening to her music and smoking water pipe till about five in the morning! In a way, and although they had completely different trajectories, I always felt there was an analogy between Ahmed and Oum Kalthoum. Both rose from the popular Egyptian culture and human warmth of their upbringing to reach the acmes of international recognition, splendor, and excellence. Both were deeply attached to Egypt and both became Statespersons and the best ambassadors their country could ever have. There was an Egyptian saying in the 1960s: "After the Lady, the President," meaning that Oum Kalthoum surpassed President Nasser in popularity. In the case of Ahmed and President Mubarak, the saying would be "After the Scholar, the President … way behind!"

Ahmed's popularity in the Arab world surpassed that of cinema or music stars. He was already a public figure prior to getting the Nobel Prize, but his fame and popularity grew exponentially thereafter. Even the jargon of ultrafast science had transpired to the laymen in Egypt, where it was not uncommon to hear a taxi driver say to a customer: "Take you there in a femtosecond!" Ahmed also felt he had the duty to help his country of origin, Egypt, and more generally of Arab societies and those of southern countries, what he called the "have-nots." This concern was high on his agenda ever since we got to know each other and it was one of our most frequent subject of conversation. His point was that one does not need much to create momentum for the improvement of Arab societies and that Science should be one of the drivers. The Arab world had all the necessary resources to rise to Excellence, and action should be taken now, he argued. His dream was to build a world-class University in Science and Technology in Egypt, which would serve as a center and a seed in the Arab world. He argued that even in Western countries, only a handful of Institutions ensures the international scientific visibility of a country and represents a model for other universities to improve and rise to a better position. His project of a University in this part of the world went through several ups and downs, in part due to political uncertainties, but his dream became true with the inauguration of the Zewail City for Science and Technology shortly before his passing. In our conversations, Ahmed stressed the need for a fully independent institution whose agenda would be Excellence. The results were not long to appear. Till the last minute, I know how much care he took to make sure things run smoothly, while being busy with the many other commitments he had and, most of all, struggling against his illness.

Despite his international fame and stature as a statesman, Ahmed remained deeply genial in his relationship with others, irrespective of their rank, origin, or profession. I witnessed its extent in 2000 when he visited the University of Lausanne as guest professor and we went together to Paris for a talk he was to give at the Institute du Monde Arabe. Everywhere we would go, he would be recognized by passers-by of Middle Eastern origin, who would stop and shake hands with him or even hug him, and he would take the time to return the greetings.

One of the last occasions I saw him was in Hamburg during the Femtochemistry Conference in 2015; despite Ahmed looking physically affected by his illness, we spent a wonderful afternoon talking about Science and Life and smoking cigars (quite a few!). He was intellectually as sharp as ever and very curious to hear about the various developments in ultrafast X-ray techniques. With his high standards for precision, he wanted to have a slide, for later use, whose contents we prepared together, which shows the landmark achievements in time-resolved X-ray science applied to chemistry, biology, and materials science. This slide is shown in Figure 10.2.

"Ahmed died at 70 but lived 140 years," said his Caltech colleague Peter Dervan. He could not find a better way of summarizing Ahmed's intense and rich life: a world scientist, a Statesman, a loving husband and father of four, a builder of generations of scientists, and at the same time a person of immense kindness and feelings for the others (see Figure 10.3).

Figure 10.2. Ahmed giving his speech after receiving the Honorary Degree of the École Polytechnique Fédérale de Lausanne in September 2009. I like this picture because the background suggests Ahmed's spirit that still accompanies many of us.

He once told me, laughing about it, how in the late 1970s, he had a very smart postdoc who turned up in his office one day to announce that he was quitting the job. Ahmed was surprised and asked him why, given that he was doing well? The other replied: "Now I understand how the Egyptians built the Pyramids."

Indeed, Ahmed was a builder of Pyramids of Wisdom. His legacy will be remembered for many generations.

Rest in Peace Ahmed, we miss you immensely.

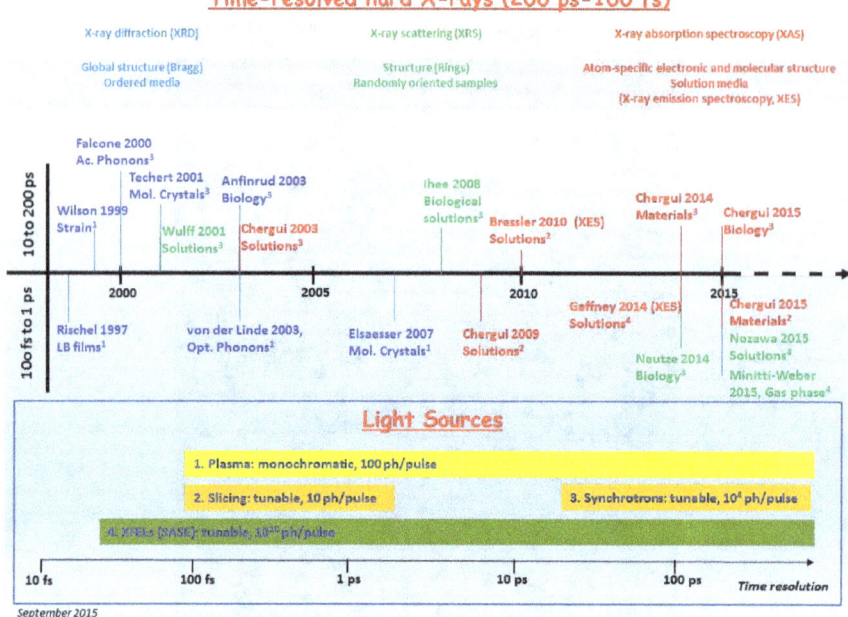

Figure 10.3. Evolution of ultrafast hard X-ray methods in diffraction (blue), scattering (green) and spectroscopy (red) and their applications to chemical and biological systems and to materials science.

References

1. J. Manz and L. Wöste, *Femtosecond Chemistry* (VCH, Weinheim, New York, 1995).
2. M. Chergui, *Femtochemistry: Ultrafast Chemical and Physical Processes in Molecular Systems*, Lausanne, Switzerland, September 4–8, 1995 (World Scientific, Singapore; River Edge, NJ, 1996).
3. J.M. Thomas, "Femtosecond Diffraction," *Nature* **351**, 694 (1991).
4. J.C. Williamson and A.H. Zewail, "Structural Femtochemistry: Experimental Methodology," *Proc. Natl. Acad. Sci. USA* **88**, 5021 (1991).
5. H. Ihee, V.A. Lobastov, U.M. Gomez, B.M. Goodson, R. Srinivasan, C.Y. Ruan and A.H. Zewail, "Direct Imaging of Transient Molecular Structures with Ultrafast Diffraction," *Science* **291**, 458 (2001).

6. A. Zewail and J.M. Thomas, *4D Electron Microscopy Imaging in Space and Time* (Imperial College Press, London, 2010).
7. M. Chergui and A.H. Zewail, "Electron and X-Ray Methods of Ultrafast Structural Dynamics: Advances and Applications," *ChemPhysChem* **10**, 28 (2009).
8. A. Zewail, *Physical Biology: From Atoms to Medicine* (Imperial College Press, London, 2008).
9. M. Chergui, Physical Biology: From Atoms to Medicine. A. Zewail (Ed.), Book Review, *Angew. Chem. Int. Ed.* **48**, 3014–3016 (2009).

Author Biography

Majed Chergui is director of the Lausanne Centre for Ultrafast Science (LACUS) at the École Polytechnique Fédérale de Lausanne (EPFL). He was trained as a physicist in London, Paris, and Berlin, before becoming Chair Professor at the Université de Lausanne (Switzerland) in 1993, and then ten years later moved to the EPFL. He is known for developing several new ultrafast spectroscopic methods that he uses for the study of the photoinduced dynamics in chemical and biological systems and in materials science. In particular, he pioneered ultrafast X-ray spectroscopy in the picosecond, then the femtosecond time domain. He also greatly improved ultrafast deep-ultraviolet spectroscopy, and pioneered two-dimensional deep-UV spectroscopy. His works have been recognized by awards such as the Kuwait Prize for Physics, the Humboldt Research Prize, the Earle Plyler Prize of the American Physical Society, and the Edward Stern Prize for Outstanding Achievements in X-ray Spectroscopy. He is also Fellow of the Royal Society of Chemistry (UK), the European Physical Society, the American Physical Society, and the Optical Society (USA). He was Editor-in-Chief of *Chemical Physics* and is now the founding Editor-in-Chief of *Structural Dynamics*, the first journal fully dedicated to ultrafast electron and X-ray based studies in chemistry, biology, and materials science.

Chapter 11

A Remembrance of Ahmed Zewail

*Chad A. Mirkin**

Although Ahmed Zewail and I were not close friends in the social sense of the term, we had a tremendous mutual respect for one another and became close colleagues while serving on PCAST, the President's Council of Advisors on Science and Technology, during the Obama Administration. Prior to that, I had never met Ahmed but knew of his extraordinary accomplishments and stature through the scientific literature and lore. I entered my first term of service on PCAST with Ahmed in the spring of 2009, and to be honest, I did not know what to expect (see Figure 11.1). The hard-driving, Nobel-Prize-winning scientist from Caltech had a reputation, among some, as someone with strong, unwavering opinions, and I was concerned it might be challenging to work with him. I could not have been more wrong.

In Washington, D.C., Ahmed and I were immersed in a policy environment that was new to both of us, and each of us was trying to figure out ways we could be useful in spurring on new science policy initiatives that would be beneficial to our country. We were surrounded by some of the most accomplished scientists, engineers, and tech/business icons in the world, and they were not deferential to our ideas — indeed, they

---

*chadnano@northwestern.edu

Figure 11.1. Ahmed Zewail sitting to the left of President Obama at a PCAST meeting in the White House.

challenged us, every step of the way, at the highest levels. I recall walking over to Ahmed at the conclusion of our first PCAST meeting and introducing myself, "Professor Zewail, it is a pleasure to meet you, but I must admit, you are awfully quiet and not living up to your reputation." He replied, "The pleasure is mine, young man. This is new territory for me, and I am just trying to get the lay of the land." His response was somewhat surprising to me. He characterized, in a very honest way, what we were both feeling — an unusual and uncomfortable uncertainty. Although both of us were quite confident in our scientific capabilities and accomplishments, how could we make comparable and significant impacts in the policy game? As with other areas, Ahmed was a fast learner. He rapidly made important contributions, and in an Obama Administration effort to effect diplomacy through unconventional means, he became President Obama's first Special Envoy for Science to the Middle East. He was proud to use his extraordinary reputation and accomplishments to try to improve relations between the US and a part of the world that was in intense turmoil.

Figure 11.2. Ahmed Zewail and Chad Mirkin in 2012 at the annual IIN symposium in Evanston, Illinois.

Throughout our service on PCAST, Ahmed and I would share rides to and from the National Academy of Sciences (the meeting venue for PCAST) and, in the process, we shared many stories, mostly about recent advances in our own labs. I grew to know him as the gentleman that he was, who had an unmatched passion for science and a constant glimmer in his eye as he talked about his current areas of interest. I eventually invited him to Northwestern University (NU) to give a plenary talk at our 2012 International Institute for Nanotechnology (IIN) Symposium, which is the largest science event at NU every year (see Figure 11.2). I expected him to give a "Greatest Hits"-type lecture. Instead, he laid out the ground work for a whole new field of 4D microscopy — the combining of ultrafast spectroscopy with 3D electron microscopy. That day, I felt like I had just heard the foundational work for a second Nobel Prize. He made his argument as a seasoned scientist, but with the enthusiasm and passion of a young assistant professor. At dinner afterward, he surprised me, telling me that he had followed my group's work on nanocluster synthesis and shape control and exclaiming that he could see ways that the new techniques he was developing could provide mechanistic insight not possible by conventional means. Ahmed personally initiated a collaboration with my group that we started shortly before his death. Sadly, he will not be able to see the outcome of that work, but his students and mine continue to explore ways of carrying this work forward.

Ahmed's legacy is remarkable. He and his work have created an indelible mark on the field of chemistry, and his contributions to the US and his native country of Egypt are a rarity. The way he carried himself was admirable, and he was model for the rest of us. He will be missed, but never forgotten.

Author Biography

Chad A. Mirkin is the Director of the International Institute for Nanotechnology and the George B. Rathmann Professor of Chemistry, Chemical and Biological Engineering, Biomedical Engineering, Materials Science & Engineering, and Medicine at Northwestern University. He is a chemist and a world-renowned nanoscience expert, who is known for his discovery and development of spherical nucleic acids (SNAs) and SNA-based biodetection and therapeutic schemes, Dip-Pen Nanolithography (DPN) and related cantilever-free nanopatterning methodologies, On-Wire Lithography (OWL), and Co-Axial Lithography (COAL), and contributions to supramolecular chemistry and nanoparticle synthesis. He is the author of over 700 manuscripts and over 1,000 patent applications worldwide (over 300 issued), and he is the founder of multiple companies, including Nanosphere, AuraSense, Exicure, and TERA-print, which are commercializing nanotechnology applications in the life sciences, biomedicine, and semiconductor industries. Mirkin has been recognized with over 120 national and international awards, including the RUSNANOPRIZE, the Dan David Prize, and the inaugural Sackler Prize in Convergence Research. He was an eight-year Member of the President's Council of Advisors on Science & Technology (Obama Administration), and one of very few scientists to be elected to all three US National Academies. He is also a Fellow of the American Academy of Arts and Sciences and the National Academy of Inventors, among others. Mirkin has served on the Editorial Advisory Boards of over 20 scholarly journals, including *JACS*, *Angew. Chem.*, and *Adv. Mater.*; at present, he is an Associate Editor of *JACS*. He is the founding editor of the journal *Small*, and he has co-edited multiple bestselling books. Mirkin holds a B.S. degree from Dickinson College (1986, elected into Phi Beta Kappa) and a Ph.D. degree from the Penn. State University (1989). He was an NSF Postdoctoral Fellow at the MIT prior to becoming a Professor at Northwestern University in 1991.

Chapter 12

4D Electron Tomography: Some Recollections of the Summer of 2000

Wolfgang Baumeister* and Juergen Plitzko[†]

In 2000, one of the authors (WB) of this tribute to Ahmed Zewail spent a few months at Caltech after being offered the Thomas Everhart Chair of Biology and Physics. Soon after arriving in Pasadena, he got an invitation from Ahmed to discuss potential collaborative projects. They met in his grand office in the Noyes building decorated with numerous memorabilia from a distinguished career. To WB's surprise, Ahmed was remarkably well informed about the authors' electron tomography work,[1] possibly through his close interactions with Sir John Meurig Thomas. John had developed a keen interest in electron tomography himself and used it with great success in the characterization of diverse inorganic materials, in particular solid catalysts.[2] Not surprisingly, given Ahmed's interest in four-dimensional electron microscopy the conversation soon centered on the issue of adding the fourth dimension to electron tomography and the potential of such a method for biology.[3] Several meetings of the author

---

* baumeist@biochem.mpg.de
[†] plitzko@biochem.mpg.de

with Ahmed and his collaborators followed over time, in which they pondered over possible strategies to make his vision a reality.

The challenges were obvious: To perform biologically meaningful experiments, the samples — molecules or cells — had to be kept in a liquid-hydrated ambience. For obvious reasons, the vacuum of electron microscopes is not well suited to perform experiments with aqueous samples. Don Parsons, then at the Roswell Park Memorial Institute in Buffalo, had built a differentially pumped hydration chamber (Figure 12.1) and demonstrated in a pioneering experiment that catalase crystals diffracted to high resolution when kept fully hydrated.[4] He presented his work at a seminal small workshop held at the Cavendish Laboratory in Cambridge in 1972. Inspired by Parson's work, I built a hydration chamber myself, copying their design, and tried to use it for studying the structure of model membranes — with very little success. The hydration chamber was not only cumbersome to use, I was also struggling with radiation damage; my carefully designed membranes vanished before I was able to acquire images. Low-dose devices and methods had not yet been developed in the early 1970s. The use of hydration chambers was soon superseded by using cold-stages, designed originally to mitigate the effects of radiation damage. Bob Glaeser and Ken Taylor in Berkeley used them to repeat the Parson's experiment but now with a frozen-hydrated catalase sample, and they attained the same resolution but with a much more practical experimental setup (Figure 12.1).[5] Their experiment heralded the beginning of cryo-electron microscopy (cryo-EM) as we know it today, a method that has become tremendously successful and is nowadays the method of choice for a wide range of cellular and molecular structural biology problems.

Another problem in working with liquid hydrated samples is radiation damage. Nevertheless, in recent years, we have seen a renewed interest in studying biological materials in liquid environments using new technologies, such as the use of silicon nitride or graphene sealed hydration chambers.[6] But the results so far are rather non-spectacular and the problem of radiation damage persists. It has been claimed by some practitioners of liquid-hydrated electron microscopy that radiation damage in liquid water may be less severe than in vitreous ice, but that is in contradiction to a whole body of experimental data from radiation chemistry and biology.[7,8]

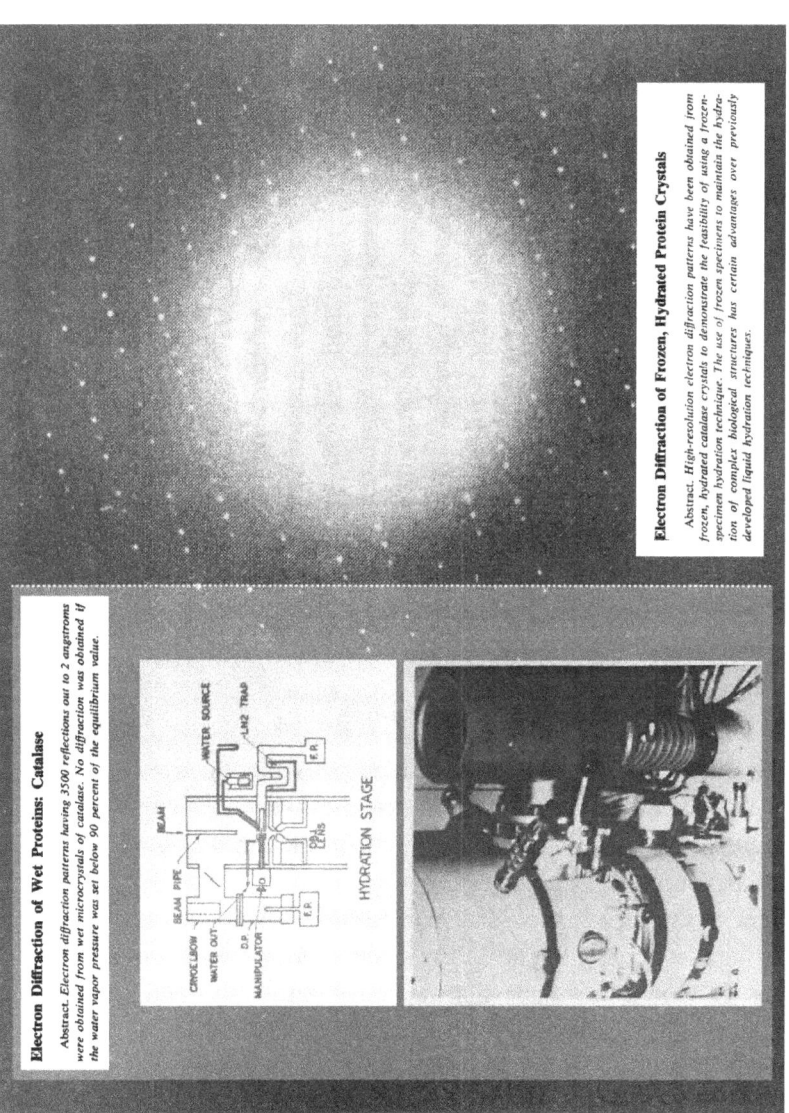

Figure 12.1. Left, schematic of Parson's hydration chamber and its external appearance. Right, electron diffraction pattern of a catalase crystal, which was frozen in liquid nitrogen.

Source: Adapted from Refs. 4 and 5.

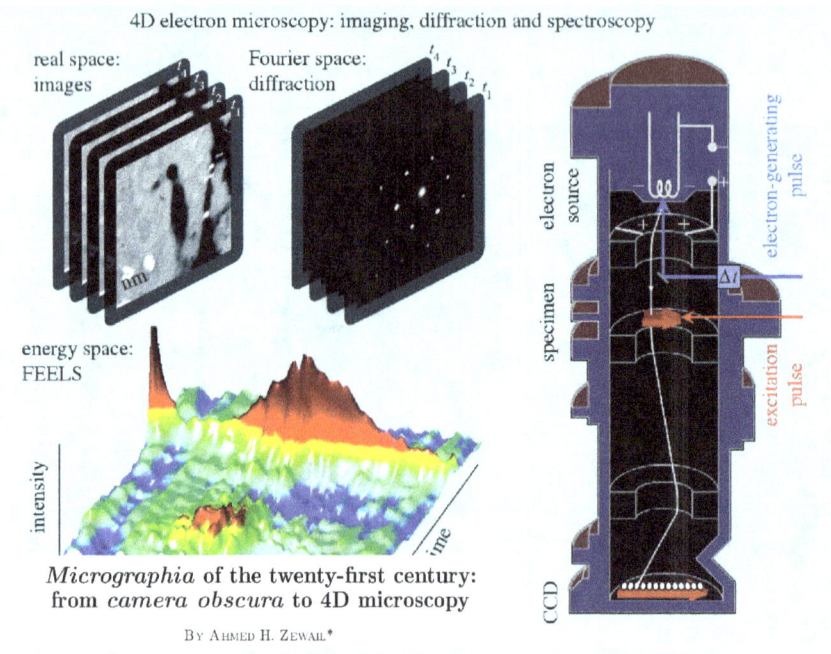

Figure 12.2. 4D electron imaging in real, Fourier, and energy spaces. The conceptual design of Caltech's UEM-2 is shown on the right.

Source: Adapted from Ref. 12.

In frozen samples, the free radicals generated by the electron beam are restricted in their movement and therefore less likely to recombine in a destructive manner. There is general agreement though that it will not be possible to perform live cell imaging.[9]

Another question we discussed with Ahmed at some length was the issue of being fast enough to outrun Brownian motion that otherwise blurs the images or tomograms such that no meaningful information can be obtained. To tackle the short timescales of this jittery motion, one has to be extremely fast, either in illuminating or in taking a picture. Ahmed belonged to the category of the "Illuminati" and he mastered the approach of extreme "short exposures" with timed pulses of electrons to obtain impressive "motion pictures".[10] Others, including ourselves, looked into the concepts of digital cameras, fast enough to capture real-time movies of the processes under investigation. However, movie making was not our primary goal, since the life in cryo-EM undeniably is frozen in time. The main motive was the improvement of the detection efficiency.

Ahmed and his scholars did reach sub-picosecond time resolution by keeping tabs on the dispatched single electron-wave packets (Figure 12.2). For him, the detector's temporal response was irrelevant. To date, the readout of direct detectors range from several tens to several hundred frames per second, and cameras with a frame rate of 1,000 frames per second or more will become available in the near future. This new generation of electron cameras allows detecting individual electron events and furthermore motion. Motion that is induced by the electron beam, though unwanted and if not corrected for, blurs the image and reduces the spatial resolution. Direct electron detection has changed the perspectives of cryo-EM in a fundamental way and there is plenty of room for further improvements.[11] However, a camera read-out speed sufficient to "catch a glimpse" of molecules jiggling around is at the moment far from reality.

Ahmed's vision of applying 4D electron tomography to biological objects is unlikely to be realized any time soon. But technological developments since 2000 hold promise to bring us closer to that aim. It is a typical high-risk, high-reward endeavor. But as Enrico Fermi once said: "An experiment with an 80 per cent chance of success of working as predicted is hardly worth doing. You should try something you don't know will work. We need someone to support the wildest and most creative ideas."

References

1. A.J. Koster, R. Grimm, D. Typke, R. Hegerl, A. Stoschek, J. Walz, and W. Baumeister, "Perspectives of Molecular and Cellular Electron Tomography," *J. Struct. Biol.* **120**(3), 276–308 (1997).
2. P.A. Midgley, E.P.W. Ward, A.B. Hungría, and J. Meurig Thomas, "Nanotomography in the Chemical, Biological and Materials Sciences," *Chem. Soc. Rev.* **36**, 1477–1494 (2007).
3. A.H. Zewail, "Four-Dimensional Electron Microscopy," *Science* **328**, 187–193 (2010).
4. V.R. Matricardi, R.C. Moretz, and D.F. Parsons, "Electron Diffraction of Wet Proteins: Catalase," *Science* **177**, 268–270 (1972).

5. K.A. Taylor and R.M. Glaeser, "Electron Diffraction of Frozen, Hydrated Protein Crystals," *Science* **186**, 1036–1037 (1974).
6. W. Timp and P. Matsudaira P, "Electron Microscopy of Hydrated Samples," *Methods Cell Biol.* **89**, 391–407 (2008).
7. U.M. Mirsaidov, H. Zheng, Y. Casana, and P. Matsudaira, "Imaging Protein Structure in Water at 2.7 nm Resolution by Transmission Electron Microscopy," *Biophys. J.* **102** (4), L15–L17 (2012).
8. R.M. Glaeser, "Electron Microscopy of Biological Specimens in Liquid Water," *Biophys. J.* **103**(1), 163–164 (2012).
9. N. de Jonge and D.B. Peckys, "Live Cell Electron Microscopy is Probably Impossible," *ACS Nano* **10**(10), 9061–9063 (2016).
10. J.S. Baskin, H. Liu, and A.H. Zewail, "4D Multiple-cathode Ultrafast Electron Microscopy," *Proc. Natl. Acad. Sci. USA* **111**(29), 10479–10484 (2014).
11. D. Agard, Y. Cheng, R.M. Glaeser, and S. Subramaniam, "Single-Particle Cryo-Electron Microscopy (Cryo-EM): Progress, Challenges, and Perspectives for Further Improvement," Chapter 2 in *Advances in Imaging and Electron Physics*, P.W. Hawkes (Ed.) (Elsevier, Amsterdam, 2014), pp. 113–137.
12. A.H. Zewail, "Micrographia of the Twenty-First Century: From Camera Obscura to 4D Microscopy," *Philos. Trans. R. Soc. A* **368**, 1191–1204 (2010).

Author Biography

Wolfgang Baumeister studied biology, chemistry, and physics at the Universities of Muenster and Bonn, Germany, and he obtained his Ph.D. from the University of Düsseldorf in 1973. From 1973 to 1980, he was Research Associate in the Department of Biophysics at the University of Düsseldorf. He held a Heisenberg Fellowship spending time at the Cavendish Laboratory in Cambridge, England. In 1982, he became a Group Leader at the Max-Planck-Institute of Biochemistry in Martinsried, Germany, and in 1988 Director and Head of the Department of Structural Biology. He is also an Honorary Professor of the Physics Faculty at the Technical University in Munich. In 2000, he spent time at the California Institute of Technology as a Moore Distinguished Scholar. Wolfgang Baumeister made seminal contributions to our understanding of the structure and function of the cellular machinery of protein degradation, in particular the proteasome. Moreover, he pioneered the development of cryo-electron tomography. His contributions to science were recognized by numerous awards, including the Otto Warburg Medal, the Schleiden Medal, the John M. Cowley Medal, the Louis–Jeantet Prize for Medicine, the Stein and Moore Award, the Harvey Prize in Science and Technology, and the Ernst Schering Prize. He is a member of several academies, including the US National Academy of Sciences and the American Academy of Arts and Sciences.

Juergen Plitzko studied mineralogy and physics at the University of Tuebingen and obtained his Ph.D. degree in Chemistry from the University of Stuttgart in 1998 while working at the Max-Planck-Institute of Metals Research, Stuttgart. He was a Postdoc at the Lawrence Livermore National Laboratory, Livermore, CA, USA, and at the Max-Planck-Institute of Biochemistry. In 2012, he was appointed full Professor at the Bijvoet Center for Biomolecular Research, Utrecht University, The Netherlands, and consulting Group leader at CEITEC, Central European Institute of Technology, at the Masaryk University in Brno, Czech Republic. In 2016, Juergen Plitzko became a Group leader for cryo-electron microscopy (cryo-EM) at the Max-Planck-Institute of Biochemistry in Martinsried, Germany. Juergen Plitzko's general research objective is the development and application of innovative tools and technologies in cryo-EM and more specifically in cryo-electron tomography (cryo-ET). His work paved the way to understand structure–function correlations of macromolecular complexes in their native cellular context. His activities, such as the recent development of focused-ion-beam micromachining for frozen-hydrated samples had an important impact on bridging the gap between molecular and cellular structural biology. His contributions have clearly stimulated cryo-ET of eukaryotic cells and tissues and have been recognized in 2015 with the Ernst–Ruska award.

Chapter 13

Anatomy of a Friendship and Collaboration

*John Meurig Thomas**

It is often proclaimed that a stylist is someone who does and says things in memorable ways. From an analysis of his experimental prowess, written contributions, lectures, and even the details of the illustrations he used in his published papers or during his lectures to scientific and other audiences, Ahmed Zewail, by this or any other definition, was a stylist *par excellence*.

For more than a quarter of a century, I interacted with Ahmed (and members of his family) very regularly. Sometimes he and I spoke several times a week through long-distance calls. Despite our totally different backgrounds, we became the strongest of friends, and we got on with one another like the proverbial house on fire. We collaborated scientifically and adjudicated one another's work as well as that of others. We frequently exchanged culturally interesting stories. We each relished the challenge of delivering popular lectures. In common with very many others, I deem him to be unforgettable, for a variety of different reasons. He was one of the intellectually ablest persons I have ever met. He possessed elemental energy. He executed a succession of brilliant experiments. And,

---

*jmt2@cam.ac.uk

almost single-handedly, he created the subject of femtochemistry, with all its magnificent manifestations and ramifications.

From the time we first began to exchange ideas, I felt a growing affinity for his personality and attitude. This was reinforced when I told him that, ever since I was a teenager, I had developed a deep interest in Egyptology and a love for modern Egypt. On several occasions during my visits to Cairo, we were able to spend time together in that magical city. At first, from the early 2000s onwards, we were often accompanied by a lady who was a professor of chemistry at the American University in Cairo (AUC), Jehane Ragai, who later (in 2010) became my wife. Only a few years ago, Dema, Jehane, Ahmed, and I had dinner together in Cairo, at about the time when the constructional work for the Zewail City of Science, Technology and Medicine was getting underway. Happy days!

I welcome this opportunity to sing Ahmed's praises; and I begin by drawing attention to his Nobel Lecture: "Atomic Scale Dynamics of the Chemical Bond using Ultrafast Lasers." The reprint from *Lex Prix Nobel 1999* consists of 100 pages of extraordinarily beautiful vignettes and illustrations: there are 36 Figures, most of them consisting of color illustrations that could be incorporated, without change, into standard university texts. The whole article is redolent of deep scholarship: it mingles historical appreciation with frontier scientific achievements; and some of the portrayals in his article, typified by Figure 13.1, are worthy of being used as tapestries on the walls of museums of science and other temples of scholarship.

The version of his Nobel Lecture that appeared in *Angewandte Chemie International Edition* is the largest ever such Nobel article: it spanned some 43 pages. The excitement, background debts to early collaborators (e.g., Robin M. Hochstrasser, Charles B. Harris of Berkeley, Richard B. Bernstein, John C. Polanyi, Fred C. Anson, Rudolph A. Marcus, John D. Roberts, Peter Dervan, Jackie Barton, Harry Gray, Vince McKoy, and all his collaborators at Caltech and elsewhere) as well as the doubts and triumphs associated with frontier research, are beautifully chronicled in this *magnum opus*.

Two years after winning the Nobel Prize, Ahmed, along with his able colleague, Spencer Baskin, published a summarizing account of his work, entitled "Freezing Atoms in Motion: Principles of Femtochemistry and

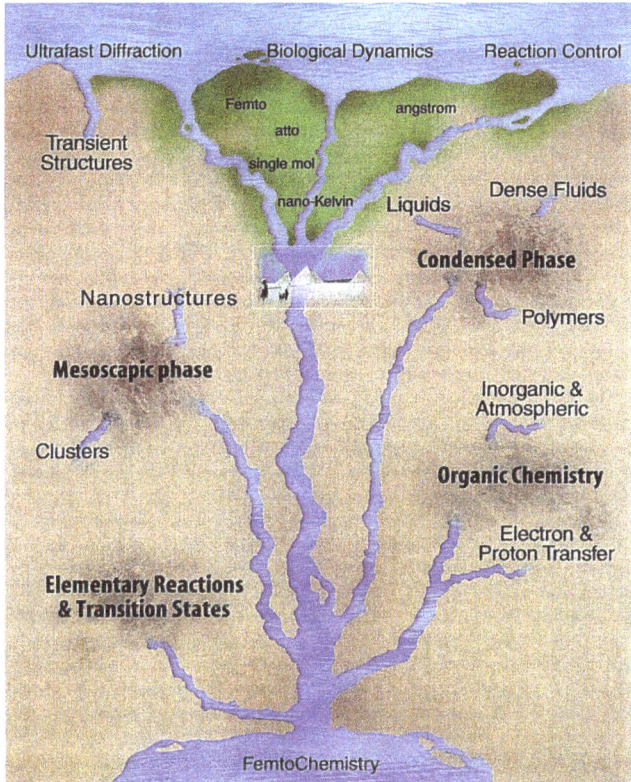

Figure 13.1. This is Figure 20 of Zewail's Nobel Lecture: *Femtochemistry*, reprinted from *Les Prix Nobel, 1999*, p. 162.

Demonstration of Laser Stroboscopy"[1] that was targeted at pedagogs. It exhibits the same exhilarating attributes as his Nobel Lecture, but the illustrations are now presented in a manner that is immediately usable by high school teachers.

Having had countless conversations with him, and having written a book together, as well as a recent research article, I feel impelled to enumerate briefly the qualities that I admired in Ahmed. These include his:

- Prodigality of output;
- His general celerity of action;
- The technical virtuosity of his experimental skills;

- The profundity of thought of his theoretical excursions;
- His efficiency in mentoring students, scholars, and visiting scientists;
- His remarkable enterprising ventures in fundraising, especially for the establishment of Zewail City, during the course of which he mobilized the energies of many of his fellow countrymen.

Ahmed was also deeply committed to alleviating the plight of the "have-nots" not only in his native Egypt, but elsewhere in the Middle and Far East. He worked hard for the cause of the underprivileged, the underdogs, and especially the millions of children worldwide who receive no education.

I first got to know Ahmed in Santa Barbara at the Molecular Crystals International Symposium in 1977, to which we had each been invited to give plenary lectures by Mostafa El-Sayed. At that event, Ahmed disclosed his strategy for tackling coherence in molecular and crystals systems. In retrospect, we can now see that those early endeavours by the young Caltech Assistant Professor — without tenure at that time — constituted the first steps in his rectilinear path to Stockholm in 1999. (Of my lecture at that Symposium, Ahmed was later to write: *"To this day, I can recall the way John presented his work and particularly the way he handled the Chair of his session. In a pre-emptive strike designed to secure more time for himself he said: 'Mr Chairman, I am about to finish,' meaning he needed another five minutes or more!"*) It was not until 1990, at a Royal Society Discussion Meeting in London on "Fast Reactions," that I had the opportunity of speaking again to Ahmed. I had already read many of his publications, which clearly indicated that he was well on the way to interrogate transition states in chemical conversions involving the rupture and creation of bonds — work that had already elicited worldwide acclaim.

I engineered a private discussion with him at that Royal Society meeting and I showed him the photograph presented in Figure 13.2. I pointed out to him that, in 1826, Faraday had initiated a series of Friday Evening Discourses at the Royal Institution (RI), at which eminent men and women of science gave popular lectures to intelligent lay audiences. I also drew to his attention that Faraday, Maxwell, J. J. Thomson, Rayleigh, and Rutherford had given Friday Evening Discourses there. Would he be

Anatomy of a Friendship and Collaboration 91

Figure 13.2. Michael Faraday delivering the Royal Institution Christmas Lectures in 1856, with Prince Albert in the chair (seated directly opposite him).

willing to come and give us an account of his work? I also told him that George Ellery Hale of Caltech had given a Discourse on solar vortices and their magnetic effects in 1904; that R. A. Millikan had spoken on cosmic rays in the 1930s; that Linus Pauling, in 1948, had described the nature of the forces between biological macromolecules; and that yet another Caltech scientist, Murray Gell-Mann had spoken on elementary particles in the 1960s.

To convince him further that he should accept an invitation to perform, I also mentioned that other famous Americans, such as Margaret Mead, A. H. Compton, Roald Hoffmann, and Edwin Hubble, had given Discourses at the Royal Institution.

Ahmed agreed. And, in early 1991, it was a pleasure to entertain Dema and Ahmed in the Director's Flat for dinner before whisking him away to relax in solitary confinement in the lecturer's room prior to his appearance in the Lecture Theatre. (Faraday believed that, before delivering any important lecture, the speaker should relax on his or her own in a quiet room for at least 15 minutes. That tradition is still maintained.)

When Ahmed entered the darkened theatre, he was overwhelmed by the magnitude of the audience. The auditorium was, in English parlance, jam-packed. His opening words were, "This large crowd must be under the impression that the Egyptian speaker tonight was to be Omar Sharif!" He proceeded to give an enthralling account of his work, in such charming and exhilarating terms that he brought members of his audience almost to the brink of ecstasy. The first slide that Ahmed showed that night transported the audience to ancient Egypt and reminded them that Egypt is the "cradle of civilization." Humankind's first efforts in art, agriculture, architecture, astronomy, medicine, and dentistry, not to mention civil constructions like temples, obelisks, statues, and pyramids, were made on the shores of the Nile. His slide (Figure 13.3) had the image of Akhenaten (the father of monotheism) alongside the beautiful picture of the sun's rays emanating from the sun god, Atum. Ahmed said, "This is the first known image that depicts that light travels in a straight line."

Also in 1991, Ahmed and his colleague, Williamson, published in *PNAS* an extremely important paper on femtosecond diffraction. The

Figure 13.3. The first slide shown by Ahmed Zewail at his Friday Evening Discourse at the Royal Institution, April 1991.

News and Views editor of *Nature* invited me to write a commentary on it, which I did in June of that year. The Williamson–Zewail paper was revolutionary, since it paved the way to dynamic electron microscopy. In my conclusion, I said that this work at Caltech was likely to lead to the dawn of a new era in crystallography and microscopy. And indeed it has done so. Zewail's 4D Electron Microscopy (4D EM) has transformed the whole corpus of physical, biological, medical, and engineering science, bringing forth a succession of results of unprecedented importance and beauty. Zewail's 4D EM work has enhanced time-resolution by ten orders of magnitude. It is little wonder that the Nobel Laureate, Roger Kornberg, in commenting on Zewail's most recent book, *The 4D Visualization of Matter* describes it as "a chronicle of an extraordinary journey of invention and discovery."

In 2006, the American Philosophical Society at Philadelphia celebrated the Tercentenary of the birth of Benjamin Franklin. Ahmed's lecture on "Franklin's Vision" on that occasion was a masterpiece. It encapsulates much of his philosophy and presages his later crucial work as an emissary for President Obama. I reproduce his opening paragraphs below:

> "ON THIS SPECIAL OCCASION of the Tercentenary, I am especially delighted to speak in honor of a polymath and an American icon, Benjamin Franklin. Since his death in 1790, Franklin has been revered, memorialized, and made into an educational, financial, and political icon. Through his collective work this sage has climbed to the apex of human endeavor in the sciences, public service, and statesmanship in international relations. Such great heights for a man of wit and wisdom are reached by very few in the world, both then and now.
>
> I have real connections to Franklin, though not biological in nature. My first science home in America, the University of Pennsylvania, was founded by him; the pre-Nobel recognition I received from the Franklin Institute as the medal in his name; and election to this August Society has indeed strengthened my bonds to Franklin's home of knowledge and to Franklinian ideals of "promoting useful knowledge." In my office I have his bust for a daily reminder of what it means to be a scientist in service of society and a citizen of the world at large.
>
> For me personally, Franklin is a hero, not only for his unique and remarkable scientific contributions in the 1700s, but also for his humanitarian vision and his belief in the power of learning. He best used his own power as an accomplished scientist to influence world politics and peace. Perhaps the greatest of all of his achievements was his efforts to secure America's independence and peace with England. Today, it is Franklin's vision, with his spirit of compromise and eloquence, that we need in order to reach a dialogue and peace in our troubled world.
>
> Some attribute the origin of calamity and conflict to a clash of civilizations. I do not. Being a cultural product of both "East" and "West" with a voyage of confluence, not clash, I do not find a fundamental basis for the so-called "clash of civilizations." What is important is to emphasize cooperation, not confrontation, and to understand that we live in an interdependent "flat world" (in the words of Thomas Friedman) that cannot be peacefully sustained with huge disparities in wealth and

conspicuously inconsistent policies. Let me quote what Franklin said more than two centuries ago in Poor Richard's Almanac:

Who is wise? He that learns from everyone.
Who is powerful? He that governs his passion.
Who is rich? He that is content.

These words radiate vision, thought, and humor. Franklin added, "Who is that? Nobody." True, perhaps, but an important point is that being rich and powerful has meaning and responsibility, and that hegemony, if we are wise and learn from history, does not work in the end."

But this is not the "Franklin's vision" that I will be discussing in the remaining time. Rather, I would like to ask how, at the atomic and molecular level, did Franklin actually see?

Ahmed then proceeded to give a brilliant account, intelligible to lay folk of the atomic and molecular events involved in vision.

As is well known, Ahmed was eternally indebted to his early education and upbringing in Alexandria, Egypt. By a happy coincidence at the Franklin Tercentenary, there was another famous scientist-mathematician of Arab stock: Sir Michael Atiyah (former President of the Royal Society and of the Royal Society of Edinburgh, Fields Medallist, and arguably one of the world's foremost mathematicians) (Figure 13.4). He was, like Ahmed, educated in his pre-university days in Alexandria and Cairo.

In December 2007, Ahmed Zewail masterminded a three-day symposium held in Cambridge to celebrate my 75th birthday. He gave one of his spectacularly gripping lectures to open the event, attended by many of Cambridge's premier scientists, Sir Brian Pippard and Archie Howie, each former Heads of the Cavendish Laboratory; Lord [Martin] Rees, President of the Royal Society; Amyand David Buckingham; Sir Colin Humphreys; Lord Jack Lewis; and most of the major figures in solid-state, materials, and surface chemistry in Europe and Asia. The lecture was another of Ahmed's greats — and it was much talked about in Cambridge in subsequent years.

Figure 13.4. Photograph taken at the Franklin Tercentenary Celebrations, American Philosophical Society, Philadelphia, April 2006.

Source: By kind permission of the American Philosophical Society.

Figure 13.5 shows Jehane Ragai sandwiched between Ahmed and me at that event. She is one of the tens of millions of Egyptians who rejoice in Ahmed's phenomenal achievements.

In the summer of 2008, I spent four whole weeks in Caltech and a week in Yosemite National Park with the Zewail family and the families of Dema's brother and sister. My purpose in doing so was to write a monograph with Ahmed on *4D Electron Microscopy: Imaging in Space and Time*, a venture that emerged partially as a result of the Robert A. Welch Symposium in Houston, held in October 2007, on "Physical Biology: From Atoms to Medicine," where I had presented a talk on "Revolutionary Developments from Atomic to Extended Structural Imaging."

Another factor that brought Ahmed and me together in this way was the set of short scientific reviews that I had earlier written, extolling the virtues of the remarkable succession of breakthroughs that the Zewail team had accomplished at that time.

Figure 13.5. Ahmed Zewail, Jehane Ragai, and the author at the University of Cambridge, December 2007.

Together in the Noyes Building, we worked with frenetic zeal and obsessive commitment. To avoid irresistible temptations — like discussions with Jack Roberts (always illuminating) — we rarely went to the Athenaeum for lunch. It was during this period of interaction that I grew to learn more about Ahmed and his intellectual stature.

Shortly before I arrived in Caltech in 2008, Ahmed had grown so excited about the prospect of our composing a monograph on 4D EM together, that, in July of that year, he faxed me a tentative draft of the content of our intended monograph. Figure 13.6 is a copy of this draft, which again illustrates Ahmed's incredible energy and vision.

In the course of my numerous discussions with Ahmed over the years, I also acquired other insights into what propelled him with such brilliant intensity:

- His love of knowledge for its own sake;
- His unique combination of patience, passion, pertinacity, and perspicacity;

> *Modern Electron Microscopy*
> *Imaging & Revolutionary Inventions*
>
> July 16, 2008
>
> I. Preface
>
> II. Principles
>   a. Electron Duality & Uncertainty (λ; ΔE; Relativistic...)
>   b. Coherence (of States, Spatial, Temporal)
>   c. Resolution & Contrast
>   d. The Microscope & Its Variants
>
> III. 2D and 3D Microscopy
>   a. High-Resolution
>   b. Tomography
>   c. Holography
>   d. Applications (from Materials to Biology)
>
> IV. 4D Microscopy & Diffraction
>   a. The Concept of Single Electron Imaging
>   b. Noisy CW vs. Coherent Pulsing
>   c. Coherence Volume: Single Electron vs. Single Pulse
>   d. Applications (Reactions, Phase Transitions...)
>   e. Potential of Biological Imaging
>
> V. Electron vs. X-ray
>   a. Real Space vs. Fourier Space
>   b. Numerical Estimates
>   c. Current State of X-ray
>
> VI. Epilogue
>
> *Dear John! Greetings! What do you think of this outline I came up with? All the best Ahmed*

Figure 13.6. Ahmed Zewail's suggested framework for our joint book, later published (in 2010) titled *4D Electron Microscopy Imaging in Space and Time* by Imperial College Press.

- His profound interest in history generally, but the history of science in particular.

My Eurocratic view of who discovered what and where, was often corrected by Ahmed, who reminded me that, for 700 years, the language of science was Arabic. He pointed out that, in Cairo, in 1000AD, the Iraqi-born Al Hazen had invented the *camera obscura* and that this Arab scientist's *The Book of Optics* greatly influenced later European scientists, such as Galileo. He also drew to my attention that, in his beloved Alexandria, Aristarchus had suggested that the earth circulates the sun some 18 centuries before Copernicus. Ahmed also recalled that Eratosthenes, the Librarian in Alexandria, proved that the earth was spherical and calculated its circumference with amazing accuracy 1,700 years before Columbus sailed on his epic voyage. Ahmed also rejoiced that it was in his native city, Alexandria, that Hero invented the principle of the jet engine (long before Frank Whittle).

While working alongside Ahmed, and during my continual interaction with him over the years, I also discovered that his knowledge of the fundamentals of physics and chemistry were exceptional. (I did not know about the Kapitza–Dirac effect — even though it originated in Cambridge — until I met Ahmed). He executed his work in an enviably dispatchful manner, and each paper or book was the product of intense contemplation and lucubration.

Looking back again at the writing of our monograph on 4D EM, I vividly recall the rapidity with which Ahmed picked up new ideas. I had been working in chemical electron microscopy for some 40 years. Concepts and ideas that took me a fair time fully to comprehend were acquired by Ahmed as swiftly as the blinking of the Biblical eye. When one contemplates the contents of his most recent monograph *4D Visualization of Matter*, it is staggering to think that he mastered so comprehensively and so expeditiously the multiple facets of "static" electron microscopy, and added to them so magnificently the time domain. I cherish the fact that, in a small way, I helped him conquer new vistas in temporal and spatial electron microscopy, and I was touched when, in his "Acknowledgments" to *4D Visualisation of Matter*, he remarked: "Besides our intellectual bonds and common interests, John was the first to see clearly the significance of the 4D imaging developments over the past decade."

During his after-dinner speech on the occasion of my 75th birthday, seated close to the German Chancellor, Angela Merkel (Figure 13.7), Ahmed said of me: "He has written many obituaries and given eulogies of distinguished scientists to salute their contributions to science and society. I have repeatedly told John to write my obituary in advance…"

There is a sad irony associated with the fact that, owing to a favorable dispensation of providence, I was, in effect, thanks to the invitation from President Rosenbaum of Caltech, able to give a verbal account of my admiration of Ahmed at the Gala Dinner held in Caltech on February 26,

Figure 13.7. By a happy coincidence, one of my former collaborators who presented a paper at my celebratory symposium was the famous computational chemist, Professor Joachim Sauer, of Humboldt University, Berlin. It was our great good fortune that his wife, the German Chancellor, Dr. Angela Merkel, was able to join us at the celebratory dinner in Peterhouse, Cambridge, on December 15, 2007.

Figure 13.8. Photograph (taken by R. van Daalen of Elsevier) at the Zewail Symposium in Boulder, Colorado, just before we each presented our talks at the ACS Meeting, March 2015.

2016, when much of what I have written above was communicated verbally in his presence.

There are numerous memories of Ahmed which will remain with me for as long as I live. One is captured in Figure 13.8. The other, rather trivial ones, are of interest in this collection of tributes. The first deals with his endearing competitive spirit. In his scientific work, his desire to be the best is largely responsible for his enduring brilliance. But, even in small things, he strove to surpass. The example I quote here pertains to the inscriptions he and I each wrote in the copy of our book on *4D Electron Microscopy* that was presented to Jehane, my wife (see Figure 13.9).

Shortly after our book appeared in December 2009, I presented Jehane with an inscribed copy. A month later, when Jehane went to a function in Cairo at which Ahmed was due to speak, she asked him to add his

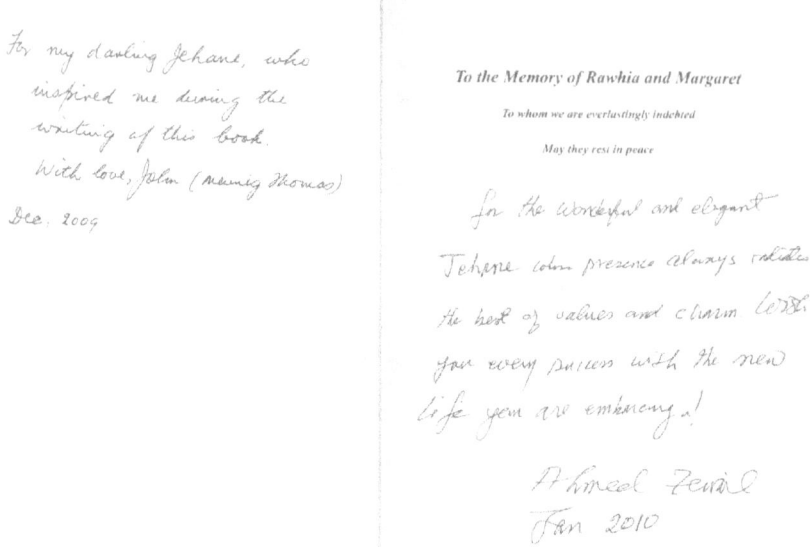

Figure 13.9. Inscription made by me (December 2009) and by Ahmed (January 2010) on the inside cover of the Zewail–Thomas monograph, 2010, on electron microscopy: "For the wonderful and elegant Jehane whose presence always radiates the best of values and charm. Wish you every success with the new life you are embracing! Ahmed Zewail, Jan 2010." I had written "To my darling Jehane, who inspired me during the writing of this book."

inscription also. He willingly obliged, but said to her jokingly: "This is both longer and composed in better English than John's inscription."

The other vivid memory I have centers on Ahmed's infectious sense of humor. We were both in an audience at an international meeting, when the speaker, an eminent scientist, started to talk about "the genius of the carbon atom." Ahmed, on hearing this, turned to me and remonstrated, "Carbon is inanimate! It cannot exhibit genius! We shall soon be talking about the genius of this chair."

Quite apart from the myriad extraordinary things that Ahmed accomplished as a scientist, educator, creator of a new City of Science, ambassador of science, statesman, and family man, he was a life-enhancing person. He was a joy to be with. Even now, less than six months after his passing, I feel tempted, as if I were in a time warp, to share with him new, mutually interesting facts — like the discovery I made recently on reading of the life of the great American Egyptologist, James Henry Breasted, that the 104th Psalm of the Hebrews shows a notable similarity to the hymn composed by the Pharaoh Akhenaten.[2] His great sense of humor and his genuine friendship to one and all will never be forgotten.

In view of the fact that his scientific and administrative work was imbued with a sense of compelling urgency, it is fitting that I should end this tribute by reverting to Ahmed's lecture on "Franklin's Vision" that he gave at the American Physical Society. Ahmed ended that talk as follows:

> "Benjamin Franklin understood the importance of time and its centrality to our lives. Of time and life he said (Poor Richard's Almanac, 1746), Dost thou love life? Then do not squander time, for that's the stuff life is made of."

References

1. A. Zewail and S. Baskin, "Freezing Atoms in Motion: Principles of Femtochemistry and Demonstration of Laser Stroboscopy," *J. Chem. Edu.* **78**, 737–751 (2001).
2. J. Abt, *American Egyptologist: The Life of James Henry Breasted and the Creation of his Oriental Institute* (University of Chicago Press, Chicago, 2011), p. 110.

Author Biography

John Meurig Thomas is a former Director of the Royal Institution (RI) of Great Britain, London, and a former Head of the Department of Physical Chemistry, University of Cambridge, and is now an Honorary Professor at the Department of Materials Science, Cambridge. He received his education at the University of Wales (Swansea) and taught and researched at that University (Bangor and Aberystwyth) for 20 years prior to joining Cambridge in 1978. His term as Director of the RI started in 1986. For over 50 years, he has researched widely in the entire domain of solid-state chemistry and is best known for his pioneering work in various kinds of chemical electron microscopy, for his contributions to crystal engineering and heterogeneous catalysis, and for transforming the study and use of zeolites and other nanoporous materials. He also elucidated the role of crystalline imperfections in governing the electronic, photochemical, and photophysical properties of organic molecular crystals. His work on single-site heterogeneous catalysts has led to many practical advances in green conversions. Following his Welch Foundation lecture in 2007 on "Revolutionary Developments from Atoms to Extended Structural Imaging," Ahmed Zewail suggested they should together produce a monograph on *4D Electron Microscopy: Imaging in Space and Time*, which appeared in 2010. A Fellow of the Royal Society (1977) and of the American Philosophical Society (1993), he is also an honorary foreign fellow of many other national academies, including the Royal Swedish Academy of Sciences; the Italian Accademia dei Lincei; the Spanish, Russian, Hungarian, and Polish Academies; and the American Academy of Arts and Sciences. For his work in heterogeneous catalysis and solid-state chemistry, he has been awarded numerous prizes, including the Linus Pauling, Kapitza, Natta, Zewail, Stokes, and Willard Gibbs Gold Medals. A new mineral, meurigite, was named in 1995 in recognition of his contributions to geochemistry, and he was knighted in 1991 for his services to chemistry and the popularization of science.

Chapter 14

Ahmed the Explorer

*Martin Pope**

In 1981, Professor Mostafa El-Sayed called me and asked whether I would be willing to present a short course, on what is now known as Organic Electronics, to a group of scientists in Alexandria. I readily agreed, and I told my wife. She asked: "Which Alexandria?" I had never thought of that. She told me to find out, and of course, he meant Alexandria, Egypt. It was there that I met Professors El-Sayed and Ahmed Zewail.

I am honored to speak in praise of the scientific contributions of Ahmed Zewail. The very choice of his name, whose initials are AZ, is prophetic; his contributions enlighten all of science. As Mathematics enlightens Physics, Physics enlightens Chemistry, and Chemistry enlightens Biology. It is all a matter of time. As one proceeds from one level to the other, the time periods for understanding the transition becomes shorter. Professor Zewail's mastery of the femtosecond makes clear the details of the passage from one field to the other, and in so doing, makes possible altering the outcome in a more desirable manner. He used his profound curiosity to elucidate the passage into and throughout Chemistry.

---

*mp3@nyu.edu

Unfortunately, his untimely death deprives us of his opening of windows into the secrets of the creation of life. Fortunately, he has shown us the way.

Sic transit Gloria Mundi.

The Discovery of Molecular Semiconductors

The discovery of molecular semiconductors marks the establishment of new fields of Physics and Chemistry and Industry, now referred to as Organic Electronics. The major types of materials involved in this field are known as conjugated organic compounds, which, in their pure state, are excellent insulators. They include individual molecules and polymers. Pope discovered unidirectional ohmic contacts to these materials that operate in the dark as well as under illumination. These insulators are thereby transformed into semiconductors. Such contacts have made it possible to use these materials for hole and electron charge injection electroluminescence, photosensitized electrophotography, photovoltaic effect, field effect transistors, biosensors, transponders, and essentially every function previously performed by inorganic devices, all without the use of toxic or expensive materials. Pope's discovery of photosensitized charge injection into organic insulators stimulated the development of dye-sensitized injection into inorganic insulators such as TiO_2, now important in solar energy utilization, such as in the electrolysis of water to produce hydrogen as a pollution-free fuel, and as a solar energy storage medium. Pope and his group discovered electroluminescence, a photovoltaic effect, and efficient exciton fission; the latter has already almost doubled the efficiency of photovoltaic devices. Pope also showed that the intrinsic photogeneration of carriers does not involve a direct band to band transition, but is preceded by a coulombic bound precursor (such as a polaron). He also co-discovered that the low intrinsic photoionization efficiency in these organic media is due to geminate recombination. Pope's discoveries have already been utilized in industrial devices such as electrophotography, organic light emitting diodes (OLED), in TV and cell phone screens, and in low-cost, easily manufactured radiofrequency identity (RFID) tags. Pope was awarded the Davy Medal in 2006 by the Royal Society of London for his discoveries.

Chapter 15

Ahmed Zewail: Advancing Chemistry

Norbert D. Dittrich*

Ahmed Zewail was a towering intellect who combined a consuming curiosity about life with a prodigious appetite for work. He also could light up a room with his infectious smile.

Ahmed's legacy ranges from his many contributions to scientific discovery to his passionate championship of science on a global scale. Through The Welch Foundation, an organization dedicated to supporting basic research in chemistry for more than 60 years, I experienced both his keen scientific insights and his work to promote the value of scientific knowledge.

His first connection with The Welch Foundation dates back more than two decades. In 1994, Ahmed was a presenter at the Foundation's 38th conference on chemical research, exploring the "Chemical Dynamics of Transient Species." His presentation, "Transient Species at Femtosecond Resolution," was enthusiastically received.

I had the pleasure of spending more time with him when the Foundation named him recipient of the Welch Award in Chemistry in 1997 for his work in establishing the field of femtochemistry (Figure 15.1).

---

* dittrich@welch1.org

Then in 2002, he was invited to join The Welch Foundation's Scientific Advisory Board (Figures 15.2 and 15.3). In that role, he was an invaluable contributor to the Foundation's mission of advancing chemistry to improve life. He, too, strongly believed in the importance of supporting basic scientific research and the long-term value it brings to society.

In addition to providing guidance on Foundation activities, from bestowing scientific awards to candidates to research grant proposals, Dr. Zewail was generous with his time in regard to our annual conference, designed to share the latest advances in a particular field.

In 2007 (Figure 15.4), Dr. Zewail chaired his first Welch's conference, "Physical Biology — From Atoms to Cells," and the following year, presented "4D Microscopy — Visualizing Materials and Biological

Figure 15.1. Ahmed Zewail celebrates with his family upon receiving the 1997 Welch Award in Chemistry.

Figure 15.2. Ahmed Zewail commemorates the 50th anniversary of The Welch Foundation in 2003 with Scientific Board members Peter Dervan, Bill Lipscomb, and Yuan Lee along with Jackie Barton.

Figure 15.3. Ahmed Zewail served on The Welch Foundation's Scientific Advisory Board for 15 years. This photo was taken in 2003, his first full year on the board.

Figure 15.4. Ahmed Zewail was actively involved with Welch Foundation's annual conferences on chemical research for more than 20 years. He is pictured here with President Norbert Dittrich and Scientific Advisory Board Chair Jim Kinsey in 2007 during a conference he chaired.

Function" at the 52nd conference titled "Biological Macromolecules: From Structure to Function."

During the last year of his life, Ahmed leveraged his famed efficiency and global network to fashion another outstanding conference that showcased his most recent passion: deciphering the behavior of complex systems from materials to biological cells through the direct visualization of structures and their evolution. While his death preceded the conference itself, his spirit imbued the presentations on exciting new research using 4D imaging, with ultrafast electron diffraction, crystallography, and microscopy, which holds amazing promise for new breakthroughs.

Dr. Zewail's work with the Foundation is emblematic of his drive to expand the role of science in the world. While only here for 70 years, Dr. Zewail made every moment meaningful, packing several lifetimes of achievement into a few decades. Sadly, Ahmed's time among us has run out. But his legacy in science will live on. The Welch Foundation salutes his memory.

Author Biography

Norbert D. Dittrich has worked at The Robert A. Welch Foundation since 1977 and has been president since 1993. He received his B.B.A. with Honors at The University of Texas at Austin in 1974. Since then, he has participated in many business and public services organizations. Currently, he is on the investment committees for the American Chemical Society and the Chemical Heritage Foundation. He also has been a member of the Foundation Financial Officers Group since 1996 and currently is a board member. He also has been a member of Philanthropy Southwest since 1983 and was president in 1997. He is on the Industry and Community Affiliates Committee for The Academy of Medicine, Engineering & Science of Texas, and the Advisory Council for The University of Texas at Austin, College of Natural Sciences Foundation. A few among many past positions in business and public service include Nonprofit Organizations Institute at UT School of Law and member of the Board of Directors and Chair at St. Agnes Academy and St. Agnes Academy Foundation. He was also a member of the Pastoral and Finance Council with Saint Cecilia Catholic Church.

Chapter 16

Four Decades in the Sub-basement — Walks of Life with Ahmed Zewail

*Spencer Baskin**

To those familiar with my career, there is no need to point out that the role of Ahmed Zewail in my life has been a profound one. Our association began 32 years ago, and I have since spent large parts of four decades in his research group in the sub-basement of Caltech's Noyes Laboratory, a tenure providing me with a different perspective from that of other friends and collaborators. In describing our relationship, the theme of "family" was and is a natural one. He expressed this theme himself, in his gracious remarks at my wedding reception, which are fortunately recorded to preserve that very special occasion in crystal clarity, though the gist of his words under those circumstances were in any case unforgettable for me. Ahmed began by saying "I probably know Spencer more than his parents. We have been married together for 18 years and it has been a very happy marriage. Spencer is almost like my son... or my brother...." While this may be viewed as an example of Ahmed's natural eloquence in crafting remarks appropriate to any occasion, for he was often very generous with

---

*baskin@caltech.edu

Figure 16.1. Dr. Ahmed Zewail at the author's wedding reception.
Photo Courtesy: Xuemei Wang.

his kind words of praise, it also expresses a truth about our relationship that I know I have felt deeply over the last months. Of course, for a complete picture of Ahmed's remarks, those who knew him will not be surprised that he also made use of the opportunity to exercise his sharp wit, repeatedly leaving the reception audience laughing heartily, as shown in Figure 16.1, possibly with his warning to my new wife: "Spencer is charming, but you don't argue with him too much."

To provide some background of our relationship, I have to go back to the early 1980s. After a stint of high school teaching in Africa, I returned to the US to look for work. It was at that point that one of my physics professors during undergraduate days at Georgia Tech suggested that I could be involved in interesting research and get paid for it by pursuing graduate studies, a path that had not until then been on my radar at all.

I weighed my alternatives, and in due course, I arrived at Caltech as a graduate student in the interdivisional program of Applied Physics, which required a full load of coursework during the first year in residence. In the summer of 1984, with a majority of required courses completed, it was time to choose a faculty advisor for those who had not yet done so. The range of research options was broad, with members of four divisions, including Chemistry and Chemical Engineering, participating in the Applied Physics program. Among the faculty who had indicated possible interest in accepting Applied Physics graduate students into their groups that year was a young Chemistry professor named Ahmed Zewail.

Though I am by no means a chemist, I had learned about some of Ahmed's research through David Semmes, a friend and Chemistry graduate student, also from Georgia Tech, whom I had met in my first year living in Braun dormitory and who was already a member of the Zewail group. The research program in which David was engaged involved coupling picosecond lasers to molecular beam sources for time-resolved laser spectroscopy of isolated molecules. It was intriguing and sounded nothing like chemistry to me, with exciting observations of quantum beats in "large" molecules (anthracene and stilbene), and I decided to approach Ahmed about the possibility of joining the group.

At this distance in time, the details of my first meeting with Ahmed are hazy, but I immediately felt his personal warmth and his engaging personality, and went away with my decision made. I believe he also agreed immediately to let me join his group, but in any case it had to be a quick and stress-free process, because I clearly remember that when the option's graduate advisor offered to intervene to get me accepted into the group of one of the senior faculty, I thanked him but reported that I had already found a place that I felt comfortable and welcome, and that was for me of highest importance.

On the other hand, I can only speculate on what induced Ahmed to accept me so readily. The path that I had followed before arriving in his group was atypical; it did not indicate the ambition and laser-like focus on advancement which were more familiar to him both in his own career and in most students at Caltech. I naively wanted to participate in some interesting research, and I did not think much further than that. If my experiences in Africa came up in our conversation, that may have worked in my favor. I know at least that he later expressed a sincere appreciation of that phase of my career, especially since he believed strongly in the value of education for those that he referred to as the "have-nots" in writings after the Nobel Prize, when he aspired to make good use of his new-found public recognition to address societal woes. The casual understanding that we reached in that first meeting was also of a type that appealed to Ahmed's sensibilities. In later years, he frequently professed nostalgia for an earlier era of simple and straightforward relationships, and a common theme in his stories about the good old days was his regret of the growing complexity in dealing with a burgeoning bureaucracy at Caltech and in funding agencies. I believe he was above all concerned about the effect of these changes in culture on the future of science.

Whatever he felt about my unconventional attitude initially, I can say that eventually it contributed to the success of our long association. In fact, much later, in one of many meetings that Ahmed held in his office with Bengt Nordén over the years of their friendship and collaboration, I remember Ahmed candidly explaining to Bengt and I that he perceived my particular value to the group to lie in my desire solely to be able to do hands-on research, and his confidence therefore that my interests and the interests of the group were in total conformity. He felt he needed to have someone to whom he could entrust a good deal of the day-to-day business of the group while he pursued his activities on the global stage, at first purely scientifically and later diplomatically.

My acceptance into Ahmed's group gave me the opportunity to make 036 Noyes my scientific home for the next five years. In retrospect, I could scarcely imagine a better experience, one in which the sheer beauty of the experiments captivated me: the glittering of dust particles drifting within the quartz spacer tube and across the thin and brilliantly hued intercavity beam of the synchronously-pumped dye laser and the ghostly glow of fluorescence from the UV laser path across the molecular beam were literally mesmerizing in the darkened lab, and the intricate patterns in the dispersed fluorescence spectra and quantum beat transients were complex works of art, but at the same time reflective of the laws of nature and amenable to understanding. In those relatively early years, more than later, Ahmed would sometimes come down and join in or watch the experiments in progress, checking up on what was going on, but also enjoying the thrill of discovery. It was in these days of my thesis research, studying the energy flow and dynamics of simple molecules under collision-free conditions, that Ahmed probably began to recognize that I aspired to nothing higher, in fact could not conceive of anything higher, than happily performing such experiments forever.

Indeed, after graduating and initially leaving Caltech, I continued in research positions studying molecular-beam and low-pressure-gas-phase spectroscopy and dynamics. Ahmed later offered me the chance to return to his group permanently, in an arrangement that explicitly included the possibility of my moving to Germany with him if he accepted an offer of a Max Planck directorship. My experiences in the German-speaking world, and overseas in general, I assumed were important considerations, but beyond that there was obviously the established relationship of trust. I believed by that time he knew me fairly well, though that belief was tested somewhat when, before his offer to me, he had first sent me a letter suggesting that I apply for a certain open faculty position. This was frankly disappointing to me, since I took it to indicate that he did not recognize the kind of research job in which I would actually have an interest. In declining to follow his suggestion, I may have provided him just the signal he needed to convince him I would really meet his requirements for the group.

If I had perhaps thought that I might go back to the comfortable and familiar work of my graduate student days, this was not Ahmed's idea. The research agenda going forward was defined by Ahmed's scientific vision, and a big factor in this was a strategy which he expressed clearly and that I think of as his 90% rule: in seeking to understand a phenomenon, the greatest impact of one's effort is often achieved by focusing on solving 90% of a problem, since it can be exponentially more challenging to nail down the "details" of the final 10% while rarely changing the big picture. Thus Ahmed was never deterred from moving on to a new problem by the fact that there remained unresolved questions surrounding a prior subject of investigation. His interests ranged widely and his labs continually evolved to keep pace. Each of the seven lab spaces that he would ultimately have went through multiple renovations under his direction. It was repeatedly necessary for me to move on to a new lab, with new or redeployed equipment, and new experimental techniques, usually faster than my personal inclinations would have dictated. It was a challenging, stimulating, and ultimately rewarding experience.

To return to the theme of "family," there were some memorable occasions over the years that led to development of a personal relationship between Ahmed and my family in Texas. Ahmed had become acquainted with my parents on their visits to California and always offered them a warm and gracious reception when they met. Then, when Ahmed was awarded the Robert A. Welch Award in Chemistry from the Welch Foundation in 1997, to be presented at a banquet and ceremony in Houston, Texas, he generously invited my mother and me to attend as his guests. My parents knew the chairman of the Welch Foundation Scientific advisory committee, Norman Hackerman, through their mutual connections to the University of Texas and had heard high praise for the banquet from acquaintances on the Baylor Chemistry faculty, only enhancing my mother's appreciation for the special invitation, and the October banquet was a wonderful occasion for her, and me.

Two years later, Ahmed was invited to give the Gooch–Stevens Lecture at the Department of Chemistry and Biochemistry of Baylor University, a lecture series which was distinguished by the fact that the majority of its previous speakers were Nobel Prize winners. Ahmed's acceptance gave my family, including my parents and my brother's families, the opportunity to return the favor of Ahmed's previous hospitality by hosting him on his visit to Waco. Between the invitation and the date of the lecture in early 2000, Ahmed was awarded his own Nobel Prize, adding to the luster of the occasion. To complete a remarkable intertwining between Ahmed's Nobel Prize and Waco, the American ambassador to Sweden and his wife, Lyndon and Kay Olson, who hosted a reception for Ahmed in Stockholm and sat behind Dema in the Nobel Award Ceremony in December of 1999, both grew up in Waco and were longtime friends of my family.

In this same momentous period of time, we had at Caltech our own celebrations of the Nobel Prize, ending with a gala banquet at the Athenaeum on January 15, 2000, to which former group members were invited. I had a small part in the logistics for this event and the photo session of group members which followed, bringing together former and current students and postdocs from the full 23 years of Ahmed's time at

Figure 16.2. A photo-op during the occasion of celebrating Dr. Ahmed Zewail's Nobel Prize win at Caltech in January 2000. *Courtesy*: Bob Paz

Caltech. That event remains one of the highlights of my association with Ahmed, and I was surprised and honored when he invited me to sit with him, Dema, and Vince McCoy for the photo (Figure 16.2).

In the period following the Nobel Prize, Ahmed redoubled his activities and travels, but still found time to engage in personal correspondence with my father. This gave Ahmed additional resources for his humorous jabs at me, which always included my "Texan English" when we were in conflict on the wording in a manuscript. Now he could also lament my failure to match the "generosity of praise" and refinement of my parents. I had to concede that I could offer no defense for that observation, but I could and often did argue the case for Texan over Egyptian English.

A characteristic of family relations is that interactions largely take place in the privacy of home, out of the public eye. Such was the case for Ahmed and me, with most of the time we spent together transpiring in our scientific home in Noyes Laboratory in the routine of work. There were intense periods with long, daily meetings, often with lunch ordered in to keep the flow unbroken, followed by periods of weeks or months where he would be out on travel or totally engaged in projects in which I had no part. He did not generally outline the details of his travel schedule, and assumed our work would continue uninterrupted. As he often said, he expected group members to recognize that that they were really working for themselves rather than for him, because "another paper doesn't matter much to me at this point." The eagerness with which he anticipated new results and the energy he put into working on each paper did not do much to convince us of that assertion, however. He expected us to work hard because it was natural to him, but he was also sensitive to and accommodated the limits of each individual.

Another period of high drama for Ahmed and the group occurred in 2011. In late January, Ahmed was at Caltech, engaged a variety of group-related business, when mass protests against President Mubarak broke out in Tahrir Square in Cairo. On January 31, Ahmed met with the group in Noyes before traveling to Egypt to try to influence the course of events toward a peaceful transition to democracy. Rumors were rampant that he might be persuaded to run for president, and we were not at all certain of what the future held for him, or us. I assured the group of my conviction that he had too great a passion for science to ever willingly abandon it for statesmanship, but there was always the nagging possibility that he might feel compelled by circumstances and the tide of history. He was away throughout February, and on February 26, I emailed him a birthday wish and a lengthy report on the activities in the group, telling him that there was no need to respond, but that I thought he would like to know. And I *did* think that he would like to know. For the same reason, I updated him on my mother since he was aware of some ongoing serious health concerns. Within a few hours, in the middle of the night, Cairo time, he replied succinctly: "Many thanks. I am glad your mother is doing well; give her my best wishes." This response is for me an example of the natural human touch that endeared him to so many.

As it turned out, of course, he did return to Caltech early in March and went right back to work. On March 29, he accepted the Priestley Medal of the American Chemical Society at an elegant awards banquet at its meeting in Anaheim, with his senior research and administrative staff in attendance as his guests. We had been through a roller-coaster of emotions in the preceding two months, ending on a high point by sharing in yet another celebration of Ahmed's scientific achievements.

There would be low points to follow, particularly when Ahmed's illness kept him away from the lab for long periods. But he was resilient and always managed to keep research moving forward by staying in communication even when he could not come to work in his office. From early 2014, it appeared to me that he had worked his way back to shouldering his accustomed work load, which is saying a lot. I noted in response to one inquiry on his health late in 2015 that despite his illness and ongoing medical attention, "it is remarkable that he has traveled a lot this year, as you have noticed. That impresses me, since I know it would wear me out!" It is in fact a prime testimony to his passion for science that, although he went through extraordinary distractions with international affairs, diplomatic commitments, and very challenging periods of medical treatment over the last six years, sometimes keeping him out of the lab for months, it is difficult to note any gaps in his scientific production, which continued only moderately abated to the end.

The final big celebration of Ahmed's achievements was the Science and Society Conference at Caltech on February 26, 2016, in honor of his 70th birthday and 40 years at Caltech. I had the opportunity to succinctly summarize the legacy of the Zewail group in introducing a video played that evening at the Conference banquet at the Athenaeum. With his input and approval, I highlighted the things in which I believe he took the greatest pride in his scientific life: his pioneering Femtochemistry studies of the dynamics of the chemical bond, complementing the work of his Caltech forerunner, Linus Pauling, on the nature of the chemical bond, both of which were recognized by the Nobel prize; his launching of the fields of 4D Ultrafast Electron Crystallography and Microscopy with the goal of capturing images of molecules as they undergo reactions; and his mentoring of or collaboration with roughly 300 colleagues who are now spread in all corners of the globe. I can easily think of these points as representing for Ahmed the minimum essential description of his scientific career, much as Thomas Jefferson viewed three of his many achievements as paramount, giving instruction that only these were to be inscribed on his gravestone.

Ahmed was clearly exhausted by the strain of planning and organizing the Science and Society Conference, and in March announced he would be taking time off for a thorough evaluation at the City of Hope (COH). It would be natural from the outside to assume that these events would produce great anxiety in the research group. To give some sense of how Ahmed's positive attitude and powerful optimism managed to allay any fears that we may have harbored for his recovery, I share below the message that I received from him at the end of that month:

> From zewail@caltech.edu Wed Mar 30 09:14:37 2016
> Subject: Good News
>
> Dear Spencer,
>
> Good Morning!
> Good News!
>
> I will be leaving COH on Friday!
> Pharaonic genes do help!
> Best regards
> AZ

Thereafter, he returned to work once more, welcoming new members into the group, attending group meetings, working on papers, supervising shipment of surplus equipment donations to Zewail City and planning additional donations, and otherwise reassuring us of a long future in Femtoland. On my last meeting with him on July 5, 2016, before I was to leave on vacation the next morning, after discussing a paper in progress, he enthusiastically spoke of plans for continuing work upon my return with the new capabilities provided by the lasers just installed in UEM-2. For him, the best and most exciting science was still ahead of us!

The unwavering optimism and passion that Ahmed displayed in the face of challenges that could easily have crushed his spirit offer an uplifting example to those of us who knew him well. We are also the ones who need that example the most, feeling as we do the greatest void and loss created by his absence. While confronted more than most with the sad reality of that loss in my life every day, I can also count myself as very fortunate to have been afforded the distinct privilege of working so closely and so long with an extraordinary and inspiring friend for whom the sky really was the limit.

Author Biography

John Spencer Baskin is a native of Waco, Texas, and earned a B.S. degree in Physics from Georgia Tech, followed by a year of study at the Eidgenössische Technische Hochschule in Zurich, Switzerland, and three years as a high school teacher of math and physics at the Institut Kizito in Isiro, Zaire (currently the Democratic Republic of Congo). On returning to the US, he earned an M.S. in Physics from Georgia Tech, and a Ph.D. in Applied Physics from Caltech. His Ph.D. research was carried out under the direction of Prof. Ahmed Zewail from 1984 to 1989. Following five years of research split between the Research Institute of the King Fahd University of Petroleum and Minerals in Dhahran, Saudi Arabia, and the Chemistry Department of the University of Houston, he returned to Caltech and has spent the remainder of his career there as a senior member of the Zewail group. He has worked on a variety of projects carried out in nine different labs spanning most of the techniques utilized by Prof. Zewail in his research. He has been a co-author with Ahmed and (to date) 36 other members of the Zewail group for more than 50 published articles. For the last 11 years, he has been a member of the team that has collaborated in designing, building, and developing the capabilities of the second-generation laboratory for Ultrafast Electron Microscopy and is a co-inventor on two issued patents for enhancements to the technique. He is currently working with the remaining members of the Zewail group to complete a number of the many projects that were left unfinished at the time of Prof. Zewail's death.

Chapter 17

Ahmed Zewail: An Honor to Egypt and Fellow Countrymen

Jehane Ragai*

In 1988, Ahmed Zewail arrived for the first time at the American University in Cairo (AUC). He had been invited by the Science Department, as a Distinguished Visiting Professor, and news had spread that he had made an outstanding breakthrough in chemistry. I was, at the time, an associate professor in the Chemistry Unit and was very keen, as were all members of the Department, to hear Ahmed speak, as he was due to deliver a few lectures on his latest achievements. We were not to be disappointed, as we all listened, with awe and disbelief, at his description of what was thought to be the impossible!

Ahmed explained how his new technique, often described as the world's fastest camera, uses laser pulses of very short femtosecond duration — 1 femtosecond is a millionth of a billionth of a second — to observe the ultrafast transformations and dynamics in chemical reactions.

He introduced us to his early experiment, which involved the monitoring of the process of bond breakage in (ICN* $\to$ I + CN), and we were

---

*jehaner7@gmail.com

made aware of the importance of the concept of *coherence* and its key role in observing, the dynamics at atomic resolution.

We all became very excited at the idea that a new field of Physical Chemistry, "Femtochemistry," had been introduced by Ahmed, and that the elusive and ephemeral transition states could now finally be viewed in real time. This would be the beginning of a long relationship between Ahmed and AUC, a relationship marked by honors, prizes, and friendships.

Ahmed would revisit AUC in 1989 for a few days. While receiving The King Faisal International Prize for the Sciences, he had just met in Saudi Arabia, the lovely Dema (fate had it that Dema's father was receiving the same prize for literature). In a cordial informal meeting with the Science Department, also in 1989, Ahmed alluded to the happy news that he intended, that same year, to marry Dema, this lovely and wonderful medical doctor he had very recently met.

In 1993, it would not take very long for the committee responsible for Honorary degrees at AUC, to which I also happened to be a member, to realize the importance of Ahmed's work and to award him a degree Honoris Causa. AUC would later pride itself at being the first university in Egypt to award Ahmed an honorary degree *prior* to his receiving the Nobel Prize in 1999. Ahmed was also chosen, by AUC's prescient late president, John Gerhart, to be a member of its Board of Trustees, a few months *before* he became a Nobel laureate.

Shortly after receiving the Nobel prize in 1999, Ahmed came to Egypt, accompanied by Dema and his children. I remember the great enthusiasm and excitement resulting from their visit to AUC and to the country as a whole. To the Egyptian people, Ahmed was their icon, their hope, their hero... I also recall a colleague exclaiming after hearing Dema being interviewed on TV: "She has conquered the Egyptian people as much as her husband!"

In 2001, Ahmed established at AUC the Zewail prize, given every semester at commencement to recognize AUC honors graduates "who demonstrate extraordinary commitment to the pursuit of scientific inquiry and the affirmation of humanistic values." Whenever Ahmed happened to be in Egypt and at AUC during the presentation of this award, his appearance at graduation would be received with thundering applause from the audience. Such enthusiasm led our provost to exclaim more than once: "this is the first time a scientist is received with greater applause than a football hero!"

In 2004, as I happened to be the Chair of what had now become an autonomous chemistry department (in 2000, the different units in the AUC Science Department had become independent departments in their own right), I was approached by members of the Chemistry club, asking if I could help them invite Ahmed to come and give a general lecture, as a special important activity of the club.

When asked, Ahmed, with his usual generosity of spirit, accepted, and when he actually gave the lecture, the attendance was so unexpectedly and overwhelmingly large that screens as well as seating arrangements had to be added to the AUC Oriental Hall and Fountain Area , both adjoining the principal Ewart Memorial Hall where the lecture was taking place. The then AUC president, David Arnold described it as the most attended and successful lecture in the history of AUC.

That evening I hosted a dinner in my home in honor of Ahmed with, as attendees, amongst many others, President Arnold, Provost Sullivan, and Dr Zahi Hawas, who was at that time the Minister of State for Antiquities Affairs. Jocularly, Hawas addressed Ahmed telling him: "Ahmed, I hear you are always surrounded by pretty young ladies" to which Ahmed responded: "And you, Zahi, I hear you are always surrounded by mummies"! This exchange led to a general burst of amused laughter from all attendees, as the quick-witted Ahmed had hit the right tone!

In 2011, Ahmed received the Presidential medal from AUC that is bestowed "to people who have done extraordinary things for the advancement of Egypt and for the human progress of the peoples of the world." In AUC's long history of almost 100 years, this medal was only given eight times (Figure 17.1).

I have also had the opportunity to witness Ahmed's performance in a different capacity, as president of the international jury for Material Sciences of the *L'Oréal-UNESCO Award for Women in Science*, founded by Nobel Laureates Pierre Gilles de Gennes and Christian de Duve. As a member of the jury, I could evaluate Ahmed's approach in conducting the whole evaluation process. Ahmed was impartial, showed interest in the

Figure 17.1. Lecture by Ahmed in 2011 titled "Egypt is the Hope," expressing his desire to develop a scientific renaissance in Egypt.

members' viewpoints, strived for consensus, and was a decisive chairperson. As a good strategist, he would meet with the jury members after the work was done, seeking feedback and probing what can be done to improve the effectiveness of future jury meetings. Ahmed with his intelligence, enthusiasm, and effectiveness, embodied all the characteristics of an excellent leader .

Ahmed was a true patriot and an ardent lover of his native country who had great dreams for his beloved Egypt. He wrote numerous articles urging governments to support science and technology and to invest in education. His message was "...invest in education rather than weaponry...." At the Cairo Opera house, Ahmed made it a point to give regular general lectures to packed audiences, where he repeatedly stressed the importance of education, collaborative research, and hard work.

He was a very close friend of my husband John Meurig Thomas and would repeatedly tell me: "...I consider John [to be] more than a brother to me...." The feeling was very mutual, and when they both collaborated in writing their book on 4D electron microscopy — entailing the direct visualizations in four dimensions of space and time of various biological and other materials — John described to me how stimulating and enjoyable the whole experience was. He would exclaim "...Ahmed is one of the most intelligent people I have ever encountered...!" John would describe how they would think, discuss, and rarely disagree when putting on paper the bulk of their ideas. While they were working, the unique and unmatchable melodious voice of Oum Kalthoum could all the time be heard singing in the background. This iconic singer dubbed "The Nightingale of the Arab World," was revered by Ahmed and it became clear, when once interviewed on the Egyptian TV, that he was familiar with ALL her songs and lyrics!

It was also in 2011 that the Zewail City of Science and Technology (his lifelong dream) was inaugurated with Ahmed as its chair. In 2014, John and I visited the Technology and Research Institutes of Zewail City at the Sheikh Zayed campus in Cairo. We had discussions with a large number of researchers, who shared with us the very high-quality work they were undertaking. We then visited the various sites still under construction of the main campus of the Zewail City of Science and Technology. To our great astonishment and admiration we were told, that tirelessly day and night, Ahmed was closely monitoring from Caltech all the various developments taking place there.

His dream of science coming once again to the forefront of the Egyptian world is certainly on its way to being fulfilled today!

John and I were in constant touch with Ahmed during his illness; his resilience and courage during that phase were very impressive. He would share with us his infinite gratitude to Dema, his lovely wife's constant care, encouragement, and love during these difficult times (Figure 17.2). Thankfully, after one year of its onset, Ahmed's health seemed to have slowly reached a steady state, allowing him to resume his outstanding work and continue to closely monitor the development of the main campus of Zewail City.

On February 26, 2016, John and I travelled to Caltech to attend Ahmed's 70th birthday celebrations. He looked happy, and all the speeches given by prominent members of the Caltech Chemistry and Physics Departments reflected what an impressive and outstanding star Ahmed was in this famous institution. Toward the end of the event, however, Ahmed looked quite tired; the disease seemed to have taken its toll on him.

Figure 17.2. John, myself, Dema, and Ahmed in Cairo in January 2013, close to one month before the onset of his illness.

He would depart this life a few months later on August 2, with only one important wish: that he be buried in his beloved Egypt. His last request was fulfilled, with thousands bidding him farewell in an impressive and moving military funeral in Cairo, led by President Abdel Fatah el-Sisi, where he was laid to rest. Military funerals are normally reserved for military personnel; however, Ahmed was eligible for such a distinction as he had received *The Order of the Grand Collar of the Nile*, the highest state honor that could be bestowed on an Egyptian.

Ahmed has honored Egypt and his fellow countrymen and has left a legacy of hope for a better future in his beloved country, one that withstands the test of time and which will be passed down to future generations. His intelligence, optimism, enthusiasm, and sense of humor will be greatly missed. May his memory continue to show us all the way.

Author Biography

Jehane Ragai obtained a B.Sc. in Chemistry (magna cum laude) and an M.Sc. in Solid State Science, both from the American University in Cairo. In 1976, she received her doctoral degree from Brunel, the University of West London, UK. Since then, Dr. Ragai has been a faculty member in the Chemistry Department of the American University. She has chaired the AUC University Senate and the AUC Chemistry Department, was the director of the AUC Chemistry Graduate program, and the recipient of several AUC Trustees merit awards as well as the School of Sciences and Engineering (SSE) award for her role as Chair of the Chemistry Department and the 2013 University-Wide Best Teacher Award. Given her additional interest in Archaeological Chemistry, she was a consultant to the American Research Center in Egypt (ARCE) Sphinx project, has served on the National Committee for the Study of the Sphinx, and was the member of the Board of Governors of the ARCE for seven years (2001–2008). Dr. Ragai has been, for several years, a jury member for the L'Oreal–UNESCO Women in Science award founded by Nobel laureate Christian de Duve, chaired by Nobel laureate Ahmed Zewail and recently by Academician Christian Amatore. She has lectured extensively around the world to university and museum audiences on the scientific detection of forgery in paintings and has published in 2015, a book on this subject: *The Scientist and the Forger* (Imperial College Press).

Chapter 18
Timing with Light

*Archie Howie**

Pharaoh Akhenaten and his family adoring the Aton.

Source: https://en.wikipedia.org/wiki/Great_Hymn_to_the_Aten#/media/File:La_salle_dAkhenaten_(1356-1340_av_J.C.)_(Mus%C3%A9e_du_Caire)_(2076972086).jpg. (Photograph by: Jean-Pierre Dalbéra).

---

*ah30@cam.ac.uk

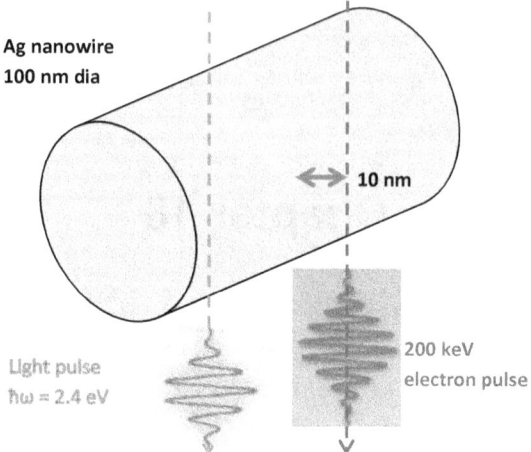

Interaction of ultrashort laser and electron pulses with a silver nanowire.

We need some stimuli which coincide
With our brief trip through life's cacophony,
Inducing neural pulses to collide
And trigger moments of epiphany.

Ahmed, that Maestro of stroboscopy
Long served as such a major stimulus;
Beyond precisely timed microscopy
He'd help us when we'd badly missed the bus!

At Cal Tech, taught by Feynman in my prime,
Great Pauling's static bond was not imbibed;
But visiting Zewail at later time
I got bond motion brilliantly described.

We came three thousand years too late to share
The shock that Akhenaten's reign displayed;
But thanks to Ahmed we are now aware
The foresight that his images conveyed.

Straight downwards from the sun the light path lies
Hands life to everyone who feels its spell.
From lying flat in profile image rise —
In three dimensions how their bodies swell!

Now Zewail's laser pulses synchronised
Can rouse electrons from the Fermi sea;
To probe a sample freshly energised
Bring four dimensions to microscopy!

How photons stretch a catalytic bond
Or drive a thin plate's complex drumming modes
Or how hot carriers in chips respond;
I'll fill this book with eulogistic odes!

Each pulse a nanowire illuminates
As triggered aloof beam electrons pass;
The near field photon pulse then operates,
Drives e-beam gains and losses to amass.

Elastic photon scattering takes place
To make a field that then electrons map;
No damage then the sample to deface
Can PINEM still fresh information tap?

All this was published swiftly on the go
Matching the shortened time he had to spend
No tedious referee the flood could slow
Or rapture of posterity suspend!

Author Biography

Archie Howie graduated with a B.Sc. in Physics at the University of Edinburgh in the same year as the now famous Tom Kibble, following his early education at Kirkcaldy High School. During 1956–1957, he held an

ESU King George VIth Memorial Fellowship at CalTech, where, sitting at the feet of Feynman and Gellman, he obtained an M.S. degree. His good luck followed him to Cambridge, where he joined Peter Hirsch and Mike Whelan just at the point where transmission electron microscopy and the imaging of defects by the diffraction contrast mechanism were taking off. His Ph.D. research there concentrated on the development and experimental testing of the dynamical theory of electron diffraction with the necessary modifications to address the problem of imaging dislocations and other crystal defects. His subsequent career at the University of Cambridge covered several new applications of electron microscopy to semiconductors and catalysts. The development of the field emission scanning transmission electron microscope (STEM) by Crewe opened up new opportunities in imaging and spectroscopy for Howie and his research team to develop — particularly, incoherent high angle dark field imaging and spatially localized plasmon spectroscopy. As Head of the Cavendish Laboratory (1989–1998) in difficult financial times, one of Archie Howie's privileges was to preside over the centenary celebrations for the discovery of the electron. Among the many Nobel Laureates who were invited to describe their ride to fame on the back of the electron was Ahmed Zewail, making the first of several visits to stun a Cambridge audience with his remarkable achievements.

Chapter 19

Ahmed Zewail — A Towering Visionary

*Colin Humphreys**

When I think about Ahmed Zewail, the image I have in my mind is of a towering visionary, who brought together his love of Egypt, both ancient and modern, and his love of science, both ancient and modern, in a powerful fusion.

I was privileged to be given a pre-publication draft of Ahmed's book with John Meurig Thomas titled *4D Electron Microscopy: Imaging in Space and Time*, published in 2010. This wonderful book brings an extra dimension, that of time, into electron microscopy. It is also full of evidence of how Ahmed not only viewed ancient Egypt through a scientific eye but also how he saw science through an Egyptian eye. For example, Figure 1.3 on page 4 of the book, reproduced as Figure 19.1 here, is of Akhenaten and his wife Nefertiti. The caption reads: "The significance of light-life interaction as perceived more than three millennia ago, since Akhenaten and Nefertiti. Note the light's 'ray diagram' from a spherical point source, the sun." I am a great lover of ancient Egypt, which I have visited many times. I have been a member of the Egypt Exploration Society, founded in 1882, for many years. I have seen this beautiful

---

*cjh1001@cam.ac.uk

picture of Akhenaten and Nefertiti a number of times. But it has never struck me before that it depicts the first known image showing that light travels in straight lines. It took the binocular vision of Ahmed Zewail, looking at this beautiful picture both through his scientific eye and his Egyptian eye, to reveal the extra dimension that I had missed.

Ahmed's brilliant scientific achievements are very well documented in this present book honoring his memory. I would like to make a final comment on his passion for Egypt, which came through so strongly when talking with him. Anyone who has visited Egypt, and who has a love for it, knows that walking in the magnificent ancient temples and tombs somehow gets into one's bones. The world-leading ancient Egyptian civilization was so amazing. Ahmed, born and bred in Egypt, must have experienced this evidence of a glorious past more profoundly than can any non-Egyptian. His vision of Egypt once again becoming a world leader, this time in modern science, led to his founding the Zewail City of Science

Figure 19.1. Akhenaten, Nefertiti, and their children.

Source: https://upload.wikimedia.org/wikipedia/commons/4/47/Akhenaten%2C_Nefertiti_and_their_children.jpg.

and Technology in 2011. Let us hope that this City expands and flourishes as a living memorial to this most remarkable man.

The image I vividly retain in my mind of Ahmed is not just of a towering visionary, it is also of a lover of humanity with a great sense of humor, who brought an extra dimension to life and work. I will also remember him for the twinkle in his eye!

Author Biography

Colin Humphreys is Professor of Materials Science and Director of Research at the Department of Materials Science and Metallurgy, University of Cambridge, and a Fellow of Selwyn College, Cambridge. He is a Fellow of the Royal Society and a Fellow of the Royal Academy of Engineering. He founded and directs the Cambridge Centre for Gallium Nitride (GaN). He founded two spin-off companies to exploit the research of his group on low-cost LEDs for home and office lighting. The companies were acquired in February 2012 by Plessey, which is now manufacturing LEDs based on this technology at their factory in Plymouth, UK. He founded and directs the Cambridge/Rolls-Royce Centre for Advanced Materials for Aerospace. He is the Director of a new company, 2D Technologies, to exploit the research of his group on 2D materials. In his limited spare time, he writes on science and religion and is the author of *The Miracles of Exodus* (Harper Collins, 2003), which has been translated into German and Portuguese and has an audio edition, and *The Mystery of the Last Supper: Reconstructing the Final Days of Jesus* (Cambridge University Press, 2011), which has been translated into Russian, German, Portuguese, Japanese, and Greek and has a South Asian edition.

Chapter 20

A Glimpse of the Evolution of Adiabaticity

*Noel S. Hush**

Interaction

Ahmed Zewail is remembered by the international scientific community, not only as a highly distinguished physical chemistry pioneer but also as one who was the exemplar of and advocate for the value of basic curiosity-based research, the seed corn of the harvest. His extensive efforts in bringing his native Egypt into the scientific forefront, by founding and supporting the first scientific establishment of its kind, serves as a beacon for young people everywhere.

It was my great privilege to interact with Ahmed on a number of occasions, unfortunately, usually at large meetings where opportunities to talk at length were limited. One occasion which provided the opportunity for closer interaction was his invitation to me to contribute a chapter to his book *Physical Biology: From Atoms to Molecules*,[1] a title which illustrates the unusual breadth of his interests. Another was the function at which I was awarded the prize which bears his name.

---

*noel.hush@sydney.edu.au

When Ahmed and I did manage to converse at any length on scientific matters, the subject was frequently, not unnaturally, the concept of the transition state in chemical dynamics. I had followed his work in that area with great interest, particularly because the reactions which he studied were closely related to those with which I made my first acquaintance with the field. It is always of interest to look back to the genesis of ideas and see how they develop and germinate over time. Perhaps I may be permitted a brief excursion into personal history to illustrate how fundamental ideas about reaction dynamics began to be formulated in earlier days, and what are the major insights gained in the following years.

A Transition State Milieu

I left Sydney University in 1950 to move to Manchester in England at the invitation of Professor M. G. Evans, to take up a position as junior lecturer in Physical Chemistry in his Department at the University of Manchester. This Department had been restructured following the appointment in 1933 of Michael Polanyi, co-author of the Transition State theory of chemical reactions,[2] from Berlin, and Evans, former pupil, then collaborator with Polanyi[3] had just replaced him. Evans, a dynamic man with wide interests, was generally regarded as the most outstanding UK physical chemist.

The buildings there had been quite severely damaged during the war by the German bombing, but the work of the Department had quickly recovered to the point that it was regarded as the pre-eminent Physical Chemistry department in England, and almost certainly in Europe at the time.

When called on to lecture on Reaction Kinetics, I discovered in the annexe to the lecture room (one the attendant informed me had been that in which Ernest Rutherford had given his courses in earlier times) some dusty, but in otherwise perfect condition, impressive white three-dimensional plaster casts. These had inscribed contours of potential energy surfaces with the transition region clearly marked, evidently formerly Michael Polanyi's lecture adjuncts. There were surfaces with symmetrical structures, early transition states, and late transition state structures. There was even a working model of the Walden inversion with a clamp to hold it in the symmetrical planar transition-state configuration. These were very helpful in conveying basic principles in a simple visual way.

The research in the Department covered a wide range from theory to experiment to both in concert, but much of it remained infused by Polanyi's interests. He had brought from Berlin (plus a technician) equipment for his "sodium flame" studies of kinetics of gas-phase reactions of the kind

$$RX + Na \rightarrow R\bullet + NaX$$

(RX = organic halide) which served as a test bed for transition state theory and these efforts continued by, amongst others, Ernest Warhurst (who was at the time supervising the Ph.D. work of Michael Polanyi's son John). Michael Szwarc studied unimolecular reaction kinetics; Hank Skinner and Huw Pritchard worked on energetics of bond formation and valence state theory to complement the experimental kinetic studies; and Christopher Longuet-Higgins, a major authority on Molecular Orbital theory, studied the quantum-mechanical nature of potential-energy surfaces (he coined the term "crude adiabatic" as a mathematical way of thinking about "diabatic" surfaces). There was a vigorous experimental program on solution-phase oxidation-reduction processes, and much else. M.G. Evans was heavily engaged in his own extensive work, as well as tirelessly encouraging the younger members of the Department. This was thus a very inspiring atmosphere for one such as myself, interested particularly in oxidation-reduction processes and in molecular orbital theory.

The Coupled Identical Diabatic Potential Model for Adiabatic Processes

After first working with Evans on comprehensive calculations of the energetics of oxidation-reduction processes involving hydrogen peroxide and its cognate radicals and ions,[4] and work on the adiabaticity of heterogeneous electron-transfer processes closely analogous to the sodium-flame reaction,[5] I turned to the kinetics of the reactions studied by Szwarc, which included the oxidative-reductive hydrogen atom transfer analogous to dismutation of semiquinones.[6] I discussed this by analogy with the work of Polanyi and Horiuti on proton transfer kinetics.[7] This was a seminal paper published in 1935 but had remained largely unknown to the wider scientific community because it had been published in Russian in the *Soviet Journal of Electrochemistry* and did not appear in an English version until 2007. The kinetics of proton transfer between two centers were discussed in terms of identical diabatic potentials coupled by an electronic interaction term large enough to provide an adiabatic reaction pathway, with continuous change in the electronic structure from reactant to product states, whose lower surface exhibited a transition state. The variables in this approach (for coupling to a single vibration) were the lateral displacement between the potential minima resulting from this coupling, the electronic coupling of the two diabatic potentials and the overall energy change. This provided a basic understanding of the observed essentially linear dependence of proton transfer rates on the overall energy change in the reaction, known as the Brønsted coefficient. A similar approach was taken to the Szwarc's hydrogen-atom transfer systems, utilizing molecular-orbital calculations of the overall energy change. This produced a predicted correlation between rate and overall energy change that was analogous to the Brønsted coefficient for proton transfer and was consistent with Szwarc's available kinetic data. So there were now closely analogous adiabatic transition-state representations of proton and oxidative-reductive hydrogen atom transfer processes.

The London Approach to Adiabaticity

Let us now look to the earliest quantum-mechanical calculation of the mechanisms of adiabatic and non-adiabatic reactions, to the work of Fritz London. London was a colleague of Polanyi in Berlin, generating, e.g., the London–Eyring–Polanyi (LEP) reactive potential-energy surface. After inventing (with Heitler) the chemical bond in 1927,[8] in 1932, London turned to consider methods of breaking it so as to develop a general theory of chemical reaction rates.[9] He introduced the concept of diabatic surfaces, which could be coupled to produce adiabatic surfaces. This approach was developed later by Eyring and Polanyi to produce analytical LEP potential-energy surface for general triatomic molecules, and then applied by Horiuti and Polanyi to coupled identical diabatic surfaces discussed above.

One class of reactions discussed is what London termed "charge transfer" reactions. He considered specifically atomic reactions, the example chosen to illustrate the theory being the electron transfer between potassium and chlorine atoms to form the K⁺Cl⁻ ion pair. This closely parallels, in overall mechanism, the corresponding sodium-iodine reaction for which Ahmed was first able to glimpse a transition state in his femtosecond sampling dynamics experiment.[10,11] Assuming the validity of the Born–Oppenheimer approximation, London considers[9] two electronic wavefunctions, φ_a and φ_i, characterizing the atomic (K + Cl) and ionic (K⁺ + Cl⁻) states, respectively, at internuclear separation R, coupled by the electronic interaction β. This yields two system valence-bond electronic eigenfunctions:

$$\Psi_1 = \cos\omega\, \varphi_a + \sin\omega\, \varphi_i \quad (1)$$
$$\Psi_2 = -\sin\omega\, \varphi_a + \cos\omega\, \varphi_i$$

where, with appropriate approximations,

$$2\omega = \arctan[2\beta / (E_a(R) - E_i(R))]. \quad (2)$$

The potential functions for KCl so derived are illustrated in Figure 1(a), reproduced from London's original paper.[9] The potential functions of Ahmed and co-workers for the corresponding NaI reaction are reproduced in Figure 20.1(b) and are clearly analogous,[11] the only slight difference being that London envisaged a small potential-well in the excited state. Zewail went on to show how dynamics over the transition state (Figure 20.1[c]) led to observable spectra (Figure 20.1[d]), providing a great triumph for combined theory and experiment.[11]

The Born–Oppenheimer Cusp as Key to the Interplay of the Fundamental Variables

However, there is another key feature described by London in Figure 20.1(a) that related his approach directly to what we now call "transition state theory." This is the first derivative

$$\frac{d}{dR}\omega(R)$$

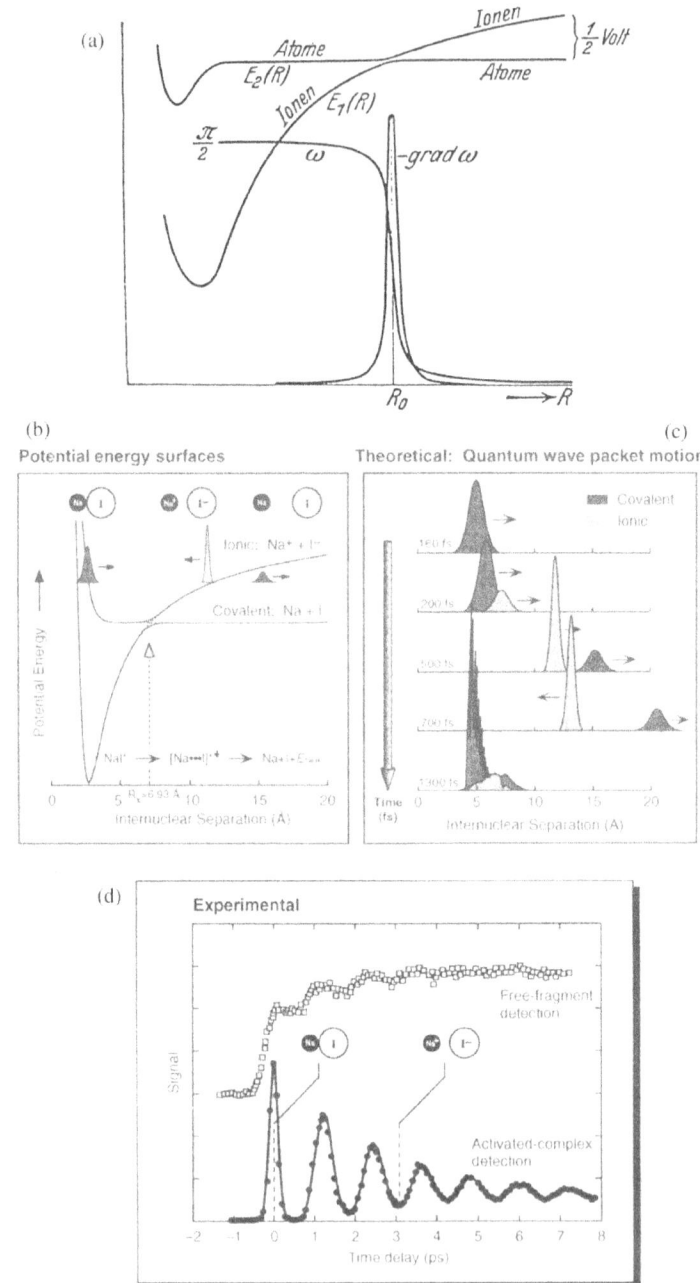

Figure 20.1. (a) London's description of the KCl reaction. (b) Zewail's description of the NaI reaction. (c) Zewail's quantum wavepacket dynamics. (d) Zewail's transition-state spectra.

which displays a long, thin spike with a cusp just below the intersection point of the uncoupled diabatic surfaces. This is the "Born–Oppenheimer cusp," and its diameter is the distance between the inflection points of the curve. It is so-called because transformation of the diabatic Hamiltonian into the adiabatic basis introduces a *pitchfork-bifurcation cusp catastrophe* that manifests itself as a derivative discontinuity in the ground-state surface when $\beta = 0$. In modern theory of dynamics of an electron linearly coupled to two (effective) harmonic potentials, critical features including the activation energy, barrier frequency, electron occupation number and hence the charge distribution at the transition state, corrections owing to Born–Oppenheimer breakdown, and electron–nuclear entanglement can be concisely expressed as functions of the cusp diameter. The reaction rate is proportional to the square of the cusp diameter. The extension to more general potential functions remains a work in progress, but London's evident appreciation of the importance of the cusp characteristics shows most remarkable insight.

The modern version (Figure 20.1[b]) of the potential landscape for the analogous NaI reaction is of course the product of work of half a century or so from initial formation and extensive development, plus enormous advances in computational resources, and the interpretation in terms of experimental quantities, in particular of wave-packet dynamics as illustrated. The process studied in both cases is an adiabatic electron transfer, which has very wide contemporary connotations.[12]

A Twist to the Tale

There is an ironic twist to the tale. While London's discussion of the mechanism is subject to more than one interpretation, it is clear from Equation (1) and from Figure 20.1(a) that he regarded the progress from one limiting state to the other as an adiabatic process, like that also drawn in Figure 20.1(b). Now in general, the distinction between adiabatic and non-adiabatic processes depends *inter alia* but critically on the distance dependence of the electron coupling term at the intersection of the diabatic potentials. London does not provided an absolute distance scale, but we can estimate it for both reactions in a simple way. We define the limiting value of $E_a(R) - R_i(r)$ as $R \to \infty$ as E_0. Since the atomic potential is essentially flat and the interaction in the ion pair is essentially Coulombic at the large distances involved, we can approximate the value R_c (in Å) at the crossing point as $14.4/E_0$, where E_0 is in eV. For Ahmed's Na-I reaction, $E_0 = 2.1$ eV and so this predicts $R_c = 6.9$ Å, close to the value (presumably for the maximum of the lower curve) shown as 6.93 Å in Figure 20.1(b), in good agreement. However, we see that London assumed E_0 for the KCl reaction to be very much less, approximately 0.5 eV, which would put the intersection point as 29 Å. On general grounds, one can safely say that adiabaticity is not possible at such a separation as β would be far too low, so that his model system does not in fact contain what we now term a transition state. Even with the currently accepted value of E_0, (0.73 eV), R_c would be also too large at 19.7 Å for a sufficient through-space interaction.

Evolution of Understanding

Of course, this is a matter of detail. What we have caught a glimpse of is the evolution of our ideas about chemical dynamics from their absolute beginning to the intricate formulations and deep experimental probings, which today mark the achievements of Ahmed Zewail. What is the driving force of such an evolution? There are two dominant current views of historians of science about the evolution of progress in science. One, that of Thomas Kuhn, stresses the sudden appearance of totally new concepts, resulting in a "paradigm shift" closing a period of "normal" science which just digests the discoveries of a previous such conceptual jump — essentially a punctuated equilibrium.[13] Another prominent view is that it is a sudden advance in tools in a particle area which initiates a new era; this has been advocated in particular by Peter Galison.[14] There are, of course, many example of this, such as the first splitting of an atom, ushering in an era of particle physics, by Cockroft and Walton with their table-top accelerator. Where does chemical reaction dynamics stand in this? Overall, London's work expands on discovery seven years earlier of quantum wave mechanics by Schrödinger, which was essentially an advance of the Kuhnian kind. Broadly speaking, Ahmed's work is essentially of the second kind. But scientists do not, in general, accept such categorizations as absolute, merely they are considered to be extrapolations to two defined limits of a process which more usually exhibits both attributes to varying degrees, and Ahmed's work is also infused by theory. He points the way ahead to even finer resolution at the atto-chemical scale. And further? "Time must have a stop," said Hotspur. This will transpire — sometime in the future.

Acknowledgments

I thank Prof. Jeffrey Reimers (Shanghai University, University of Technology Sydney) for helpful discussions.

References

1. A.H. Zewail (Ed.), *Physical Biology: From Atoms to Molecules* (Imperial College Press, London, 2008).
2. H. Eyring and M. Polanyi, "Concerning Simple Gas Reactions," *Z. Phys. Chem. Abt. B* **12**, 279 (1931).
3. M.G. Evans and M. Polanyi, "Inertia and Driving Force of Chemical Reactions," *Trans. Faraday Soc.* **34**, 11 (1938).
4. M.G. Evans, N.S. Hush, and N. Uri, "The Energetics of Reactions Involving Hydrogen Peroxide, Its Radicals, and Its Ions," *Quart. Rev. Chem. Soc.* **6**, 186 (1952).
5. M.G. Evans and N.S. Hush, "Ionogenic Reactions Involving Bond Breaking at Electrodes," *J. Chim. Phys.* **49**, C159 (1952).
6. N.S. Hush, "Quantum-mechanical Discussion of the Gas Phase Formation of Quinonedimethide Monomers," *J. Polym. Sci.* **11**, 289 (1953).
7. J. Horiuti and M. Polanyi, "Outlines of a Theory of Proton Transfer," *Acta Physicochimica U.R.S.S.* **2**, 505 (1935).
8. W. Heitler and F. London, Wechselwirkung neutraler Atome und homopolare Bindung nach der Quantenmechanik *Z. Phys.* **44**, 455 (1927).
9. F. London, "On the Theory of Non-adiabatic Chemical Reactions," *Z. Phys.* **74**, 143 (1932).
10. T.S. Rose, M.J. Rosker and A.H. Zewail, "Femtosecond Real-time Observation of Wave Packet Oscillations (Resonance) in Dissociation Reactions," *J. Chem. Phys.* **88**, 6672 (1988).
11. J.C. Polanyi and A.H. Zewail, "Direct Observation of the Transition State," *Acc. Chem. Res.* **28**, 119 (1995).

12. J.R. Reimers, L. McKemmish, R.H. McKenzie, and N.S. Hush, "A Unified Diabatic Description for Electron Transfer Reactions, Isomerization Reactions, Proton Transfer Reactions, and Aromaticity," *Phys. Chem. Chem. Phys.* **17**, 24598 (2015).
13. T.S. Kuhn, *The Structure of Scientific Revolutions* (University of Chicago Press, Chicago, 1970).
14. P. Gallison, *Image and Logic: A Material Culture of Microphysics* (University of Chicago Press, Chicago, 1997).

Author Biography

Noel Hush (b.1924) is an Australian with joint Australian–British nationality who was born and brought up in Sydney. He enrolled in the Faculty of Science at the University of Sydney in 1941; specializing in Chemistry in the later years, he was granted the B.Sc. honours degree in 1945. He was given the (then only) graduate award, the Dunlop Fellowship and commenced work toward a Master of Science degree, which was awarded in 1949. In the same year, he accepted an invitation for appointment as Junior Lecturer in Physical Chemistry at the University of Manchester, England, from the Head of the Department which had been restructured by Michael Polanyi, at that time the most renowned chemist in Britain, particularly strong in molecular orbital theory and reaction dynamics, including oxidation-reduction. In 1954, following the untimely death of Evans, he moved to the Chemistry School at Bristol University, where he set up a fruitful research group in the general area of Chemical Physics, principally theoretical but but with associated spectroscopic work, also holding numerous visiting appointments. In 1971, he returned to the University of Sydney as Foundation Professor of Theoretical Chemistry, establishing the first Department providing undergraduate to postdoctoral teaching in the subject in Australia. This flourished both in its teaching and research activities, with a nation-wide reach through its Summer Schools. As Research Emeritus Professor since 1989, he has continued to work in the area of quantum chemistry on problems of structure and dynamics with a continuing focus on electron transfer and its relation to basic features of chemical structure and reactivity and molecular electronics. He is a Fellow of the Australian Academy of Science and the Royal Society of London and is a Foreign Honorary Member of the American Academy of Arts and Sciences and of the US Academy of Sciences in Chemistry and (as affiliate) in Physics. His recognitions include the Robert A. Welch award for Chemical Science and the Ahmed Zewail Prize for Molecular Science. He married Thea Warman (now deceased) in 1950. He has two children, Julia, a medical researcher who specializes in Pain, and David, who is a classical composer.

Chapter 21

Ahmed Zewail: Science and Scientist

Joshua Jortner*

Genesis

My friendship with Ahmed Zewail originated as early as 1974, when I spent a sabbatical leave at the University of California, Berkeley, where Ahmed served as a Post-Doctoral Fellow in the group of Chuck Harris. Already at these initial stages of Ahmed Zewail's adventures in science, I was tremendously impressed by him as an outstanding, highly original, and brilliant young scientist. Ahmed attended my Berkeley lectures on intramolecular dynamics, with the participation of about 80 graduate students, post-docs, and faculty. This was an exciting endeavor, as I used to walk into the class without lecture notes, just with some reprints and preprints and talk about the metamorphosis in chemical physics, with the development of the radiationless transitions theory. Ahmed always sat in the front row, taking careful notes. When I approached him and asked him to photocopy the lecture notes for me, he was delighted to comply, commenting: "I write down everything, even your jokes." At that time Ahmed was courted by top universities in the US, and it was easy to convince my

---

*jortner@post.tau.ac.il

friends and colleagues at the University of Chicago to make an offer to this brilliant scientist. Ahmed decided to move to Caltech, where he left his mark on the development of chemical physics and related areas of science. In 1977, when I visited Caltech as a Fairchild Fellow, Ahmed had just been promoted to tenure, and was already considered as one of the leading, world-class, young scientists. He performed most original work on outstanding scientific problems, including electronic and vibrational energy relaxation and redistribution in very large "isolated" aromatic molecules, which demonstrated his deep insight in making the science community think differently about central issues. In 1977, among his many activities, Ahmed edited a special IEEE issue on laser-driven molecular dynamics. This included a paper of mine on laser isotope separation in large molecules, based on a Caltech Physics Seminar. This Seminar is famous for shaping the development of science, with the first row always being occupied by leading physicists who enjoyed making perceptive comments and interrogating the speaker. My Seminar talk triggered a hot discussion with Richard Feynman, who raised the provocative question: "Why to separate isotopes?" to which Ahmed and I responded enthusiastically.

Femtochemistry

When Ahmed started his seminal, ground-breaking work on femtochemistry, I recall an ACS meeting in 1986, which we both attended. The session on dynamics was chaired by Wilse Robinson from Texas A&M, a wonderful scientist and dear friend of both of us, who started by stating that he does not know what we will say, but it will be exciting... At that meeting Ahmed delivered one of his first talks on real-time femtosecond dissociation of triatomic molecules, with this remarkable report being followed by me addressing experimental-theoretical work on laser spectroscopy and intramolecular dynamics of jet-cooled large molecules and their van der Waals complexes. During the following years, our paths crossed at many scientific conferences, where Ahmed played a leading role as the great pioneer of femtochemistry. In 1992, when the chemical physics community celebrated the 60th birthday of Stuart Rice at the University of Chicago, Ahmed and I attended the conference in Stuart's honor, and

Ahmed preceded his impressive scientific talk with reminiscences about his relationship with the University of Chicago. In 1995, Ahmed and I attended the great scientific Lausanne Conference on Femtochemistry, under the leadership of Majed Chergui, where we enjoyed inspiring science and wonderful company (Figure 21.1). At that conference, Ahmed delivered an inspiring lecture on the history and recent progress in the field of molecular dynamics manifested by wavepacket dynamics on the time scale of nuclear motion. In 1996, Ahmed and I had the pleasure to attend the Nobel Symposium on Femtochemistry and Femtobiology in Karlskoga, Sweden. This Nobel Symposium, addressing ultrafast reaction dynamics at atomic-scale resolution, was held two years prior to the award of the Nobel Prize to Ahmed Zewail. This outstanding Symposium reflected on the remarkable scope and impact of Femtochemistry and, in particular, on the seminal role of molecular wavepacket dynamics, whose theoretical foundations were established by Joshua Jortner, Steve Berry,

Figure 21.1. Ahmed Zewail and Joshua Jortner, enjoying science and friendship at the Lausanne Femtochemistry Conference in September 1995. This photo conveys so well the superb atmosphere at that conference.

and Mordechai Bixon in 1969 and their experimental study was pioneered by Ahmed Zewail 15 years later.

Science and Values

Science is international in scope, values, and intrinsic significance. Scientists often share many exchanges of ideas regarding the role of science in shaping the future of humanity, among which the topic of science and peace plays a central role. Ahmed and I used to address the Arab–Israeli conflict, discussing ways and means for relying on international and regional scientific collaboration in promoting peace in the Middle East. I recall a dinner in 1982 which was held at a Middle Eastern restaurant in Los Angeles with Mostafa El-Sayed, a remarkable scientist and dear friend of both of us, where Ahmed, Mostafa, and I enthusiastically discussed the perspectives of the service of science in the promotion of the Egyptian–Israeli peace process.

Professor Ahmed Zewail and the Israel Science Community

Ahmed Zewail was greatly respected by the Israeli science community and by leading Israeli scientists with whom he maintained scientific and personal contacts. In 1993, Ahmed Zewail was awarded the Wolf Prize in Chemistry for his seminal contributions to the development of Femtochemistry. The Wolf Prize is an international prize in recognition of outstanding accomplishments in sciences and arts, which is given by the President of the State of Israel at a ceremony in the Knesset (Parliament). The 1993 Wolf Prize ceremony (Figure 21.2) constituted the first duty performed by President Ezer Weizmann, the newly elected President of the State of Israel (Figure 21.3). Ezer Weizmann was a great promoter of the Egyptian–Israeli peace process and was delighted to present the Wolf Prize to this world-class Egyptian scientist. About 50% of the Wolf Prize laureates win the Nobel Prize at a later stage, with Ahmed Zewail being a remarkable and notable case of this double honor. Tel-Aviv University and the entire Israeli science community honored Ahmed Zewail on May 13, 1993, when he delivered the Raymond and Beverly Sackler Distinguished Lectures in Chemistry on "Femtochemistry in Cavities" (Figure 21.4).

Figure 21.2. The recipients of the Wolf Prize, May 9, 1993, at the Israeli Parliament (Knesset) in Jerusalem. Ahmed Zewail is second from the left.

Figure 21.3. Ahmed Zewail receives the Wolf Prize from Ezer Weizmann, the President of the State of Israel on May 9, 1993, at the Israeli Parliament (Knesset). President Weizmann was a great champion of the Egyptian–Israeli peace process.

Professor Dr. Joshua Jortner
Heinemann Chair of Physical Chemistry
School of Chemistry
Raymond and Beverly Sackler
Faculty of Exact Sciences

פרופסור יהושע יורטנר
הקתדרה לכימיה פיזיקלית ע"ש היינמן
בית הספר לכימיה
הפקולטה למדעים מדויקים
ע"ש ריימונד ובברלי סאקלר

Professor Ahmed H. Zewail
The 1992/93 Distinguished Sackler Lecturer

Professor Ahmed Zewail, an outstanding scientist of Egyptian origin, serves as a Professor of chemical physics at the California Institute of Technology. He made central contributions to the modern area of ultrafast chemical dynamics, utilizing state of art lasers. He succeeded in probing, in real time, the dissociation of chemical bonds and other elementary chemical processes. His work, which opened the research area which is called femtochemistry, provides the ultimate temporal limit for the interrogation of chemical processes on the time scale of nuclear motion. Professor Zewail was awarded the Wolf Prize in chemistry for his accomplishments.

Professor Zewail visited the Chemistry Department of Tel Aviv University in 1993, had extensive discussions with the faculty members and presented a beautiful lecture on his scientific work.

קרית האוניברסיטה, רמת־אביב, תל־אביב 69978, ת.ד. 39040, טלפון: 03-6408322, פקסימיליה: 03-6415054

Figure 21.4. Report on the Sackler Lecture by Ahmed Zewail at Tel Aviv University. May 13, 1993.

The Raymond and Beverly Sackler Distinguished Lectures were established in 1980, with Rudolph A. Marcus being honored by the first Distinguished Sackler Lectureship, while in 1996, these distinguished lectures were graciously renamed by Raymond and Beverly Sackler in honor of Joshua Jortner. From 1980 until today, the Sackler–Jortner Distinguished Lectureship honored world leaders of research in the chemical sciences. Among the 30 Sackler–Jortner distinguished lecturers, five were awarded the Wolf Prize and five won the Nobel Prize, with Professor Ahmed Zewail shining within this group as a great pioneer in the Chemical Sciences.

Epilogue

The seminal scientific contributions of Professor Ahmed Zewail were recognized by the award of the Nobel Prize in Chemistry. Ahmed Zewail did not manifest the "Nobel Prize Syndrome," when some distinguished scientists "rest on their laurels" after having been honored with the Nobel Prize. Ahmed continued to explore new horizons in the chemical sciences. His outstanding contributions to Femtochemistry were followed by his great recent work on space–time and four-dimensional structural interrogation, pioneering the exploration of time-resolved structures. In this context of Ahmed moving to new and important scientific directions, we have to recall a statement made by Felix Bloch on scientific discovery. Bloch stated that a creative scientist should not only develop a single novel research field, which may be accidental, but also contribute to the establishment of two major research areas. Ahmed Zewail fulfilled the Bloch criterion in a remarkable way.

My last personal and scientific encounters with Ahmed took place in Europe and in the US. In 2003, I visited Caltech, together with my wife Ruthi, where our son Roni was a graduate student in neurobiology. It was a great pleasure to spend a day with Ahmed, discussing science and culture and talking about our families. In 2006, Ahmed and I attended a conference on Femtochemistry: Fundamental Ultrafast Processes in Chemistry, Physics and Biology in Washington D.C., under the leadership of Will Castleman Jr., where Ahmed delivered a brilliant lecture on time-resolved structures of molecules, clusters, and surfaces. My lecture on the Tel Aviv work dealing with Nuclear Fusion Driven by Cluster Coulomb explosion was attended by Ahmed, who was enthusiastic about the exploration of new and fascinating territories in ultrafast dynamics, transcending chemical dynamics toward table-top nuclear fusion in the chemical physics laboratory. This reflected on Ahmed's extremely broad and deep intensive interests in science. Ahmed Zewail was indeed a scientist without boundaries. The world science community will miss him.

Author Biography

Joshua Jortner received his Ph.D. from the Hebrew University of Jerusalem in 1960. From 1973 to 2003, he served as the Heinemann Professor of Chemistry at the School of Chemistry of Tel Aviv University in Israel. He has held visiting Professorships at the University of Chicago, the University of Copenhagen, and the University of California, Berkeley. He was the Christensen Visiting Fellow at Oxford University; served in the International Research Chair "Blaise Pascal," France; and as a Humboldt Fellow at the Humboldt University, Berlin. Jortner holds honorary doctorates from seven Universities in Israel, France, and Germany. Among his awards are the International Academy of Quantum Science Award, the Israel Prize in Exact Sciences, the Wolf Prize in Chemistry, the Honorary J. Heyrovsky Medal, the von Hofmann Medal, the Robert S. Mulliken Medal, the Joseph O. Hirschfelder Prize, the Maria Sklodowsky–Curie Medal, and the Lise Meitner Research Award of the Alexander von Humboldt Foundation. A member of the Israeli Academy of Sciences and Humanities, Jortner is a foreign honorary member of the Academies of Sciences of Denmark, Poland, Romania, Russia, India, the Netherlands, the Czech Republic, the Leopoldina National Academy of Germany, the Italian Accademia Nazionale dei Lincei, the International Academy of Quantum Molecular Sciences, the American Philosophical Society, the American Academy of Arts and Sciences, and the National Academy of Sciences of the USA. He held many honorary lectureships in Europe, Asia, US, and Israel. Jortner served as President of the Israel Academy of Sciences and Humanities (1986–1995), and as the Founding President of the Israel Science Foundation. He served as the President of the International Union of Pure and Applied Chemistry (1998–2000). His research centers on the exploration of the phenomena of energy acquisition, storage, and disposal in isolated molecules, clusters, nanostructures, condensed phases, and biophysical systems.

Chapter 22

Brief Encounters with Ahmed Zewail

*Malcolm Longair**

My interactions with Ahmed Zewail took place during a relatively short period from 2005 to 2006, but this would prove to be a crucial period for the future redevelopment of the Cavendish Laboratory — Ahmed's role was crucial. Let me fill in the background to how this came about.

I was Head of the Cavendish Laboratory, the Department of Physics of Cambridge University, from 1997 to 2005. The Cavendish had moved from the overcrowded buildings in central Cambridge, including the original Laboratory designed by James Clerk Maxwell in 1874, to a greenfield site in West Cambridge in 1974. The new Laboratory was designed at a time when the great boom in investment in scientific infrastructure in the UK was coming to an end, and Brian Pippard took the decision to design the maximum amount of floor space for the money available. The resulting CLASP design was functional with only a 25-year lifetime and was not of high quality. Among the numerous serious deficiencies were the asbestos insulation and the flat roofs which were a continuing problem with regular leaks into key experimental areas. In addition, the Laboratory was again seriously overcrowded and

---

*msl1000@cam.ac.uk

made poor use of space which could not be recovered at economic cost. In 2002, I persuaded the University that the only realistic long-term solution to the increasing problems with the CLASP buildings was to rebuild the Laboratory. We came up with the first designs for a new Laboratory in the same year, but it was a very large project indeed and so we had to plan a phased rebuild of the Laboratory.

We had to develop a strategy which was scientifically innovative and would attract philanthropic funding to be matched by the University. With Chris Dobson's arrival in the Chemistry Department in 2001 as John Humphrey Plummer Professor of Chemical and Structural Biology, we soon agreed that the Cavendish Laboratory should begin a new programme in the Physics of Biology, and two lectureships were filled in this area. Then, we planned an even wider collaboration between a number of Departments in the area of the Physics of Medicine, involving Physics, Chemistry, Biochemistry, and the Clinical School. In the end, the key relation was between Physics and the Clinical School, the Head of School Keith Peters being a particularly strong advocate of the large benefits of bringing physics and clinical medicine closer together. At about the same time, the Department received a major gift from the will of Herchel Smith to enable the establishment of chair in Physics. I took the decision that this new chair should be in the Physics of Medicine, recognizing the need to provide strong support for the new initiative as well as Herchel Smith's key role in the development of oral contraceptives.

The premier annual lecture series of the Cavendish Laboratory is named the Scott Lectures, after the donor A.W. Scott, Phillips Professor (Science) at St David's College, Lampeter in Central Wales. The series consists of three lectures given in a single week by a lecturer on a topic of current research interest. The lecturers are of the highest distinction, the first ten being Bohr (1930), Langmuir (1931), Debye (1932), Geiger (1933), Heisenberg (1934), Hevesy (1935), Appleton (1936), De Haas (1937), Siegbahn (1938), and Blackett (1939). Knowing of our initiatives in the Physics of Medicine, John Thomas strongly urged us to invite Ahmed Zewail, his friend and colleague from CalTech and Nobel Prize winner in Chemistry (1999), to deliver the 2004 series of lectures. The three lectures were in fact delivered on March 9, 10, and 11, 2005, his topic being "The Physics of Life," the lecture titles being: (1) From atoms to complexity, (2) Biological dynamics, and (3) The new dimensions.

The three lectures were a *tour-de-force* and convinced even the most hard-minded sceptical physicist of the great opportunities for innovative physics in the general area of the Physics of Medicine. Many of the technologies in, for example, femtosecond spectroscopy, were familiar to the audience, but their innovative and imaginative applications in the biomedical arena, which Ahmed had made his own, were a revelation.

I had not met Ahmed before his arrival in Cambridge, but we immediately became good friends. As he remarked in his speech during the splendid Scott Dinner held at Trinity College on Thursday, March 10, although we had not met, he recognized in a microsecond a kindred spirit. A further area of mutual interest was the pioneering book by my wife, Deborah Howard, entitled *Venice and the East: The Impact of the Islamic World on Venetian Architecture 1100-1500*. This greatly appealed to his broad cultural interests. We greatly enjoyed Ahmed's company.

By this stage, our plans for the first stage of redevelopment of the Cavendish Laboratory were well advanced, with the planning and design of a Physics of Medicine Building. An application was made to the Wolfson Foundation for a grant toward the construction of the building and, in the following months, we asked Ahmed if he would write a letter of support to the Foundation. This he did with characteristic generosity of spirit and enthusiastic support. We received the grant and with matching funds from the University, the Physics of Medicine Building was built (Figure 22.1). Appropriately, it was opened on December 16, 2008, by Sir Aaron Klug, President of the Royal Society and Nobel Prize winner in Chemistry "for his development of crystallographic electron microscopy and his structural elucidation of biologically important nucleic acid-protein complexes."

We believe we have delivered on our promises. Figure 22.2 shows an ingenious system built by Pietro Cicuta and his colleagues which enables them to observe how malaria parasites attack cells.

This was the beginning of a major drive to rebuild the whole Cavendish Laboratory through a phased rebuild of its various component parts. In November 2009, the Duke of Edinburgh performed the opening

Figure 22.1. The Physics of Medicine Building of the Cavendish Laboratory (© Cavendish Laboratory, University of Cambridge).

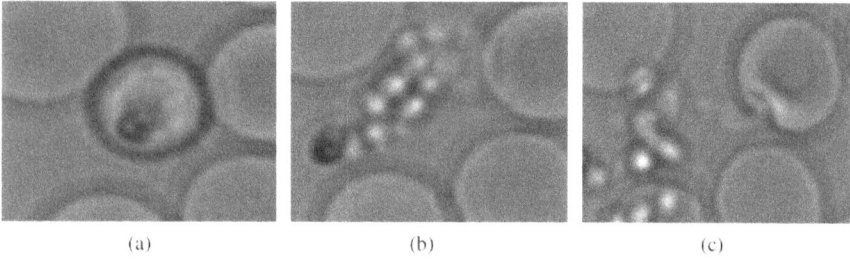

Figure 22.2. The unique automated imaging platform developed by Pietro Cicuta and his colleagues allows high-frame-rate videos of rare processes to be obtained with no human intervention. This technique has been used to observe how malaria parasites attack cells. The frames show: (a) a pre-rupture infected cell (schizont); (b) the explosive egress of the parasites, or merozoite dispersal; (c) first deformation response on merozoite-red blood cell contact (upper right cell). The piercing of the red-cell wall results in the penetration of the healthy cell by the malarial infection (*Courtesy*: Pietro Cicuta).

ceremony for the Kavli Institute for Cosmology, which brought together the cosmologists and extragalactic astronomers in the Cavendish Laboratory, the Institute of Astronomy, and the Department of Applied Mathematics and Theoretical Physics. Next the Battcock Centre for

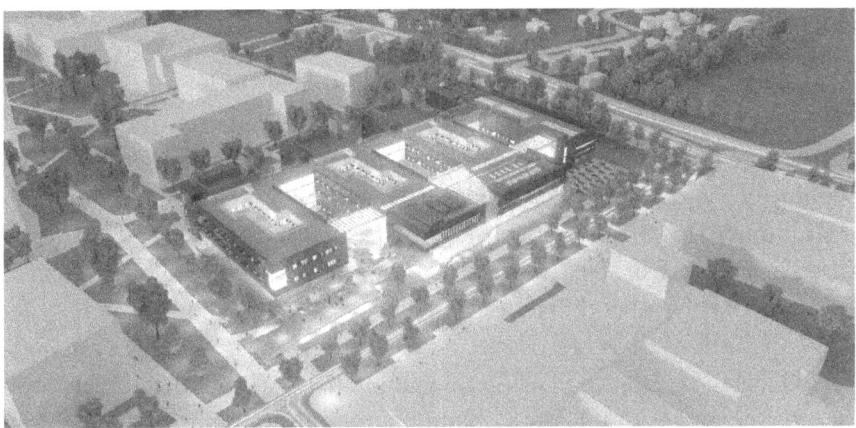

Figure 22.3. A visualization of the New Cavendish Laboratory. The research and teaching wings are connected together by parallel north-south corridors on all levels. As well as giving very good connectivity across the site, these corridors also enclose the internal courtyards, which will provide a pleasant outlook for the offices surrounding them. The research wings also have smaller courtyards set into the upper floor, which should create an excellent environment in the areas with high densities of offices.

Experimental Astrophysics was opened by the Chancellor of the University, Lord Sainsbury, in October 2013 — this building brought together the experimental and theoretical astrophysicists on the same site. In Spring 2016, the Maxwell Centre for collaboration between the Physical Sciences and Industry was opened by our generous benefactor David Harding. This building now houses a wide range of activities in the Laboratory of importance for industry. Finally, in the spring of 2017, we are in the very final stages of agreeing with the Government and the University a very large investment to rebuild the rest of the Laboratory, completing the vision of 2002. We should begin to occupy the New Cavendish Laboratory in 2021 (Figure 22.3).

But it all had to start somewhere, and Ahmed's strong support kick-started not only our major initiatives in the Physics of Medicine but also the subsequent rebuilding of the whole Laboratory. We are very grateful to Ahmed for his support, as well as his inspiration and wonderful friendship.

Author Biography

Malcolm Longair CBE FRS FRSE is Jacksonian Professor Emeritus of Natural Philosophy and Director of Development, Cavendish Laboratory, University of Cambridge. He has held many highly respected positions within the fields of physics and astronomy. He was appointed the ninth Astronomer Royal for Scotland in 1980, as well as the Regius Professor of Astronomy, University of Edinburgh, and the director of the Royal Observatory, Edinburgh. He was head of the Cavendish Laboratory from 1997 to 2005. He has served on and chaired many international committees, boards and panels, working with both NASA and the European Space Agency (ESA). He has chaired numerous committees for specific science projects, including the Planck and Euclid missions of ESA. His main research interests are in high-energy astrophysics, astrophysical cosmology, and the history of physics and astrophysics. His many books include *Theoretical Concepts in Physics* (2003), *The Cosmic Century: A History of Astrophysics and Cosmology* (2006), *Galaxy Formation* (2009), *High Energy Astrophysics* (2011), and *Quantum Concepts in Physics* (2013). His most recent book, *Maxwell's Enduring Legacy: A Scientific History of the Cavendish Laboratory*, was published in July 2016.

Chapter 23

"Peter: You Have Taught Me that Electrons are Blue!"

*Peter Edwards**

I still recall my first meeting with Ahmed in 1991 at the University of Birmingham. He was visiting our mutual friend, and colleague, Ian W.M. Smith (and the person who appointed me to my first ever Chair!). As every contributor to this marvellous volume will attest, one could not fail to be impressed by Ahmed's warm and engaging character, his ebullient personality, and his quite remarkable grasp of science — even in areas of science seemingly far beyond his own.

At Birmingham, Ahmed was also to give a presentation "Femtochemistry: The New Frontier" to a University-wide audience. Of course, his lecture was truly breath-taking; that was the first time that I had encountered his remarkable qualities as a lecturer.

In the following dinner, hosted by the then Vice-Chancellor, Sir Michael Thompson, I had the real privilege and pleasure of sitting directly opposite Ahmed. After the first few minutes exchanging pleasantries with his hosts, and in a lull in the developing conversation, Ahmed turned to me and said "Peter; I have long been intrigued by your work on metal solutions and the solvated electron — the amazing fact of the chemical

---

*peter.edwards@chem.ox.ac.uk

solvation of a fundamental particle — the electron, and your paper in Nature on trapped pairs of electrons." At that point I knew I was surely in the presence of a kindred spirit, not only someone who had opened up an entirely new frontier in the studies of molecular science but also someone who had an acute awareness of one of the most venerable areas of chemistry — the study of solutions of alkali metals dissolved in liquid ammonia.

In the next few hours, following on in Ian Smith's office afterwards (!), we discussed what Ahmed and colleagues[1] were later (in 2008) to term "... The Discovery System of Solvation." I recounted in detail my excursions in 1980 and 1981 to the Royal Institution in London to look through the laboratory notebooks of Sir Humphry Davy to follow up a hunch, triggered by my Ph.D. supervisor at Salford University, Ron Catterall, that Davy could well have seen the spectacular blue and bronze colours of electrons in metal-ammonia solutions soon after his discovery of potassium and sodium — and over half-a century before the first published observation by Weyl in 1864.

Over those late drinks in Ian Smith's office, Ahmed was genuinely excited when I showed him photographs of my discovery[2] of the entries from 1808 in Davy's actual laboratory notebooks, two of which are reproduced here as Figures 23.1 and 23.2.

A further entry is given in Figure 23.2, with the English prose much beloved by Ahmed "...an ebullition of the potassium took place, its fine silver colour became blue...it then recovered its original colour, but not its splendour."

I also showed him a photograph (Figure 23.3) of a rapidly dissolving globule of sodium-potassium alloy in liquid ammonia. He turned to me "Peter, your unearthing of Davy's (unpublished) discovery in 1808 was breath-taking, but, equally important, Peter: You have taught me that electrons are blue!" I reminded Ahmed of a comment made to me by Joshua Jortner on these spectacular colors: "Once seen, never forgotten."

Our interaction (I would hope I can say, our friendship) developed through various exchanges over the next five years. Then in 1996, I was invited by my friend and colleague, Sir Fraser Stoddart (at that time also at Birmingham) to organize the 1996 British Association for The Advancement of Science; Annual Festival of Science. Inspired by Ahmed's

"Peter: You Have Taught Me that Electrons are Blue!" **169**

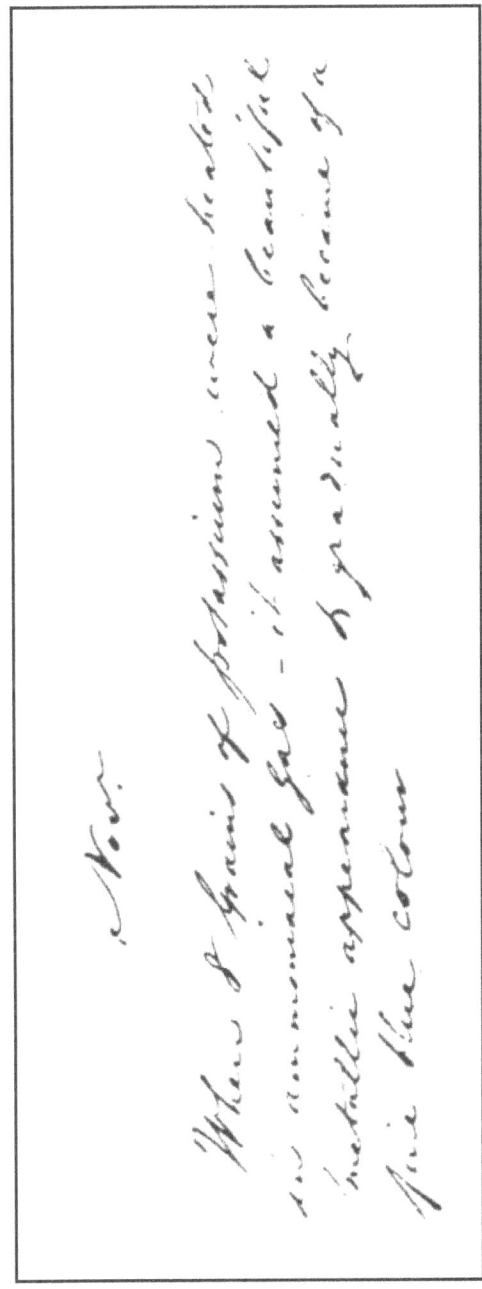

Figure 23.1. An entry from Sir Humphry Davy's laboratory notebook, November 1808.

Source: P.P. Edwards, "The Electronic Properties of Metal Solutions in Liquid Ammonia and Related Solutions," in *Advances in Inorganic Chemistry and Radiochemistry*, H.J. Eméleus and A.G. Sharpe (Eds.), pp. 135–185 (1982).

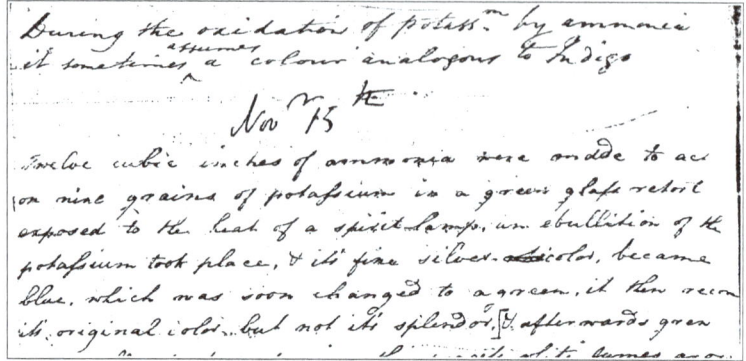

Figure 23.2. An entry from Sir Humphry Davy's laboratory notebook, November 15, 1808.

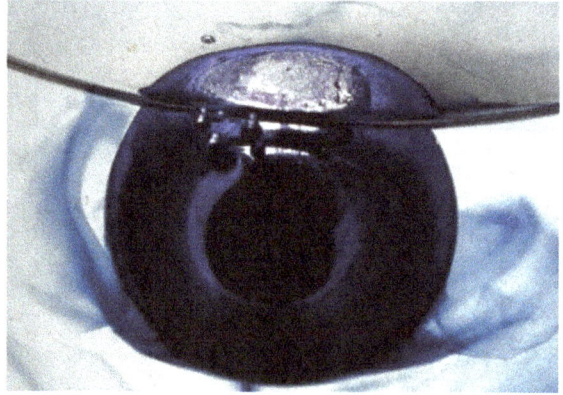

Figure 23.3. The blue solvated electron in solution.

natural and spontaneous approach to multidisciplinarity ... taking lasers out of the physicist's laboratory and onto the chemists' bench ... I decided on the theme "A Vision for Chemistry: Frontiers Across Boundaries." With the strong support of Sir Fraser Stoddart and Rob Jackson of the BAAS, and our colleagues in a most vibrant Chemistry department at Birmingham, I gathered together what I regarded, and still regard, as a truly stellar group of scientists who came to Birmingham to share our vision. The program included two Nobel Laureates, Harry Kroto and

Jean-Marie Lehn, and one (surely!) soon-to-be Nobel Laureate, Ahmed himself and, of course, our own Sir Fraser Stoddart, some 20 years hence!

When Ahmed arrived for the final day, he keenly, and instantly, asked to see the line-up and order of speakers. He rapidly perused the entire program and congratulated me of what he called a truly marvellous achievement in bringing together such a galaxy of speakers under a visionary theme that he instantly identified with.

At that stage, I had scheduled Ahmed to open the final session. After looking again at the program, he paused, and with that marvellous, mischievous smile, noted: "Peter, dear friend; I am so, so pleased that I am scheduled for the final session of your outstanding meeting. You know, Peter, I always think that the final speaker of such a prestigious conference always carries with him/her the hopes and aspirations of the organiser as the audience disperses into the wide world. What would give me infinite pleasure is for me to carry that great and pleasurable burden for you, Peter."

The (amended) final session is shown in Figures 23.4 and 23.5.

Bless you, Ahmed ... we miss you already...

A VISION FOR CHEMISTRY: FRONTIERS ACROSS BOUNDARIES

BRITISH ASSOCIATION FOR THE ADVANCEMENT OF SCIENCE, ANNUAL FESTIVAL OF SCIENCE

9-13 September 1996

School of Chemistry, The University of Birmingham, Edgbaston, Birmingham B15 2TT

The central mission of the 1996 BAAS Festival will be to celebrate - and to highlight - our subject as the forefront science that we all recognise it to be. The Festival will highlight the intellectual challenges and frontiers - "the big issues" - in our Science of Chemistry. The theme of our meeting, *Chemistry: Frontiers Across Boundaries*, was specifically chosen to reflect the pivotal, interfacial role that Chemistry now enjoys with its sister disciplines of Physics, Biology and Materials Science. A list of speakers is given below.

A View Across the Frontier

The Lord Lewis FRS (University of Cambridge)
"A Chemist's View of Measurement in the Environment"
Professor C N R Rao FRS (Indian Institute of Science, Bangalore)
"Amazing Metal Oxides"
Professor A Simon (Max Planck Institute, Stuttgart)
"A Chemist's View of the Phenomenon of Superconductivity"
Dr N K Terrett (Pfizer Limited)
"Drug Discovery by Combinatorial Synthesis"
Professor F J DiSalvo (Cornell University)
"Combinatorial Synthesis: Can it be Exploited in Solid State and Materials Chemistry?"
Professor N W Isaacs (University of Glasgow)
"How Bacteria Harvest Light for Energy"
Dr A B Holmes (University of Cambridge)
"Seeing the Light with Polymers"

The Physics and Chemistry of Interstellar Matter

Sir Harold W Kroto FRS (University of Sussex)
"Carbon in Space Comes Down to Earth"
Professor W J Nellis (Lawrence Livermore Laboratory)
"Metallization of Fluid Molecular Hydrogen at 140 GPa (1.4Mbar)"
Professor D A Williams (University College London)
"Molecules in Space"
Dr D Ward-Thompson (Royal Observatory Edinburgh)
"When a Star is Born"

Public Lectures

Professor P P Edwards FRS (University of Birmingham)
"High Temperature Superconductivity: Towards Room Temperature"
Professor I W M Smith, FRS (University of Birmingham)
"Gas-Phase Reactions at Low Temperatures: Towards Absolute Zero"

BAYS Lecture

Dr I Gameson (University of Birmingham)
"Conductors: The Good, The Bad and The Super!"

Kelvin Lecture

Dr T M McLeish (University of Leeds)
"Polymers, Peptides and Pondweed, the Simplicity of Complex Fluids"

The Periodic Table of the Elements

Dr P W Atkins (University of Oxford)
"The Periodic Kingdom"
Professor I P Grant FRS (University of Oxford)
"Relativity - The Hidden Force Behind the Periodic Table"
Professor F Hensel (University of Marburg)
"Relativity and the Properties of Gold and Mercury"

Chemistry in the Autumn at Birmingham

Ms H P Sharman OBE
Opening of Practical Laboratories, School of Chemistry
Dr R G & Mrs R Plevey (University of Birmingham)
"Chemical Magic" (Lecture Demonstration)

The Science of Small Structures: Mesoscopic Systems/Nanoscience

Professor S Mann (University of Bath)
"Biomimetic Materials Chemistry: From Magnetic Proteins to Skeletons in the Beaker"
Professor M J Kelly FRS (University of Surrey)
"Mesoscopic Systems: Will They Ever be Manufacturable?"
Professor H Ahmed (University of Cambridge)
"Single Electrons and Single Atoms in Electronics for the Next Century"
Dr S Williams (Hewlett-Packard)
"The Promise and Potential of Quantum Electronics"

Complex Systems

Professor T Kunitake (Kyushu University)
"Building Supramolecular Structures by Means of Self Assembly"
Professor W L Jorgensen (Yale University)
"Molecular Recognition"

Chemistry Presidential Address

Professor J F Stoddart FRS (University of Birmingham)
"Learning to Read and Write in Chemistry"

Nanochemistry and Femtochemistry

Professor J-M Lehn (Université Louis Pasteur, Strasbourg)
"From Molecular to Supramolecular Chemistry. Steps Towards Complexity"
Professor A H Zewail (California Institute of Technology)
"Femtochemistry - Chemists End of the Time Race"

Professor P P Edwards FRS
Tel: +44 (0)121 414 4379   Fax: +44 (0)121 414 4442

Figure 23.4. The British Association for the Advancement of Science (BAAS) Festival at Birmingham (1996).

```
A VISION FOR CHEMISTRY: FRONTIERS ACROSS BOUNDARIES

                     Friday 13 September

          Location: Haworth Lecture Theatre — Room 101
                          Morning

                      —The Kelvin Lecture—

              Chair: Dr R.S. Lehrle (Birmingham)
0930-1015     "Polymers, Peptides and Pondweed, the Simplicity of Complex Fluids"
              Dr T.M. McLeish
              University of Leeds

                 —Nanochemistry and Femtochemistry—

              Chair: Professor I.W.M. Smith, FRS (Birmingham)
1015-1100     "From Molecular to Supramolecular Chemistry: Steps Towards Complexity"
              Professor J-M. Lehn
              Université Louis Pasteur, Strasbourg

              Molecular chemistry has developed a wide range of very powerful procedures for
              building ever more complex molecules from atoms linked by covalent bonds.
              Beyond molecular chemistry lies supramolecular chemistry which aims at
              constructing highly complex systems from components held together by
              intermolecular forces.  From molecular recognition, to self-organisation, to
              programmed chemical systems it progressively leads up the ladder of complexity.

1100-1130     Coffee

1130-1215     "Femtochemistry — Chemists End of the Time Race"
              Professor A.H. Zewail
              California Institute of Technology

              Mankind, ever since the invention of photography, has been searching for better
              time resolution to directly observe the motion of objects in our universe. In the last
              half century, chemists have ended the time race. With lasers it is now possible to
              see atoms and molecules in motion in femtoseconds, a millionth of a billionth of a
              second. The principles of this new field of science and technology will be
              reviewed with examples from chemistry, biology and engineering.

1215-1230     Closing Remarks

1300-1400     General Assembly
```

Figure 23.5. The Final Session of the BAAS Festival at Birmingham (1996).

References

1. I.-R. Lee, W. Lee and A.H. Zewail, "Dynamics of Electrons in Ammonia Cages: The Discovery System of Solvation," *ChemPhysChem* **9**, 83–88 (2008).

2. P.P. Edwards, "The Electronic Properties of Metal Solutions in Liquid Ammonia and Related Solutions," *Adv. Inorg. Chem. Radiochem.* **25**, 135–185 (1982).

Author Biography

Peter P. Edwards is Professor of Inorganic Chemistry at the University of Oxford, and previously Head of Inorganic Chemistry from 2003 to 2013. His research interests include metal–insulator transitions, and future energy materials with a particular emphasis on new-generation, high-performance materials for hydrogen production and storage; carbon dioxide activation and utilization; low-cost, high-performance semiconductor thin films for solar power applications; and high-temperature superconductors. Following B.Sc. and Ph.D. degrees at Salford University, Edwards spent periods at Cornell (concurrently, British Fulbright Scholar and National Science Foundation Fellow), Cambridge (Lecturer and Director of Studies in Natural Sciences, Jesus College), as the Co-Director of the first Interdisciplinary Research Centre in the UK (that in superconductivity), and at Birmingham (Professor of Chemistry, and of Materials), before assuming his present position at Oxford in 2003. He was elected Fellow of the Royal Society in 1996 and elected to the German Academy of Sciences, Leopoldina, in 2009. In 2012, he presented The Bakerian Prize Lecture of the Royal Society, was elected to an Einstein Professorship of the Chinese Academy of Sciences, and was an International Member of the American Academy of Arts and Sciences and the American Philosophical Society. He was awarded the Armourers and Brasiers' Materials Science Venture Prize for his work on transparent conducting oxides. In 2013, he was elected to the Academia Europaea. He is Co-Founder (with T. Xiao and H. Almegren) of the King Abdulaziz City of Science and Technology–Oxford Petrochemical Research Centre (KOPRC), now designated as a Centre of Excellence in Petrochemicals.

Chapter 24

How I Lost My Funding to Zewail

*Shaul Mukamel**

I had known Ahmed since he had joined the faculty at Caltech in 1976. Our research has been strongly intertwined, and my theoretical work was greatly influenced by his pioneering studies. We had closely interacted throughout his career, starting with his picosecond studies of intramolecular vibrational redistributions (IVR), following through his femtochemistry era and finally his 4D ultrafast electron diffraction.

The field of laser chemistry was launched in 1975 by several reports of selective dissociation of SF_6 by intense CO_2 laser (frequency 1,000 cm^{-1}). The dissociation, first reported by Abratzumian Ryabov with Letokhov, was a ten-photon process. It generated great excitement and raised the hope that laser selective chemistry might be used to direct and control reactions to different products than are possible thermally; for example, the selective breaking of a stronger bond leaving the weaker one intact. One application that drew considerable scientific and government attention was the possibility of laser-induced isotope separation, leading to many schemes to achieve this goal in atoms and in molecules. In this

---

*smukamel@uci.edu

rapidly developing field, international conferences were held on almost a weekly basis, and results received extensive coverage in the regular press. I had studied and calculated possible mechanisms of selective multiphoton dissociation and how it could compete with IVR so that coherence could be maintained long enough for chemical processes to occur prior to energy thermalization. I had the opportunity in 1977 to collaborate with Nicholas Bloembergen, Eli Yablonovich, and their student Jerry Black at Harvard on the interpretation of their careful multiphoton dissociation study of SF_6. Their experiments and analysis showed that this was a thermal process: due to anharmonicities and the high density of vibrational states, IVR takes place before any bond can break, and the laser was just serving as an expensive heating device. This and many other studies caused the wide interest in laser chemistry to sharply decline. Selectivity and coherence effects were not possible by simply subjecting a molecule to an intense noisy infrared laser.

In 1978, I had collaborated with Rick Smalley at Rice, who had just set up his supersonic beam apparatus to obtain frequency-domain fluorescence measurements of cold molecules at cryogenic temperatures, allowing much better spectral resolution compared to room temperature spectra. His students were focusing on measuring the spin-orbit coupling in the methylene radical before I convinced them that IVR was a much more important issue and suggested they study it in various alkyl-substituted benzenes. Although the frequency domain data showed some indirect evidence of IVR, it was clear that only time-domain measurements could provide clear-cut and unambiguous probes for IVR. When I suggested this to Rick, he said, "Only Zewail can do it!" These experiments were indeed performed a few years later by Zewail.

When I started my faculty position at the University of Rochester, I thought it would help me obtain funding if I were to organize a workshop on IVR and quantum chaos. This was the hot topic in chemical physics at the time since it held the key for the realization of laser selective chemistry. I contacted several funding agencies who all turned me down right away, but two program directors from the Air Force Office of Scientific Research (AFOSR): Larry Davis and Larry Burggraf, were very enthusiastic, and eager to generously sponsor this conference even beyond what I had requested (this has never happened to me since!). The meeting was held in Rochester in October 1985. I invited about 30 speakers; the meeting went very well and reflected the excitement surrounding the topic and its relevance to coherent laser control of chemical reactions. Zewail, who was performing picosecond measurements in molecular beams at the time, gave a very passionate talk about "IVR and chemical reactivity," forcefully laying out the case for carrying out femtosecond (fs) experiments that could directly and unambiguously monitor vibrational motions. To achieve the fs time-resolution, he needed a new laser system. The Air Force grant officers were impressed by Zewail's talk and requested a preliminary proposal immediately. A full proposal was recommended for funding in August of 1986. Frank Wodarczyk was in charge of the Molecular Dynamics portfolio at AFOSR under which Zewail received his initial funding and was involved in the decision. He recalls, "At the time I remember being leery of providing so much money to one project, but Davis had enough foresight and faith in Zewail's ideas and in his femtochemistry proposal to commit the funding to start the project." Larry Davis had made sure that Zewail could order the equipment needed right away, and AFOSR has continued to support the femtochemistry program ever since.

The outcome of the meeting was that Zewail had received on the spot a positive response to his request for substantial AFOSR funding to build his first femtosecond system; however, when I politely inquired at the end of the meeting for possible support of my ambitious theoretical effort, I was told that no funding was left at that time (since it all went to Zewail!). In retrospect, my meeting was too successful. I take some comfort that this was my modest contribution to the birth of the field of femtochemistry.

I recall submitting a referee report to Zewail in his capacity as an editor of *Chem. Phys. Lett.* I liked the paper and indicated that it was as impressive as the pyramids, built by the Israelites in Egypt. As soon as he got the report, he called me and bitterly complained that the pyramids were in fact built solely by the Egyptians and that my statement (which was based on biblical evidence) is totally false.

Femtosecond lasers in the early 1980s opened up the way to probe photophysical processes of chemical reactions in real time. The field of laser-induced chemistry was revived with the development of femtosecond sources with stable phases, and coherent control schemes which allowed for the rational design and realization of laser-selective chemistry. I had worked on developing the density matrix, Liouville space pathway approach to nonlinear spectroscopy, recognizing how the field of ultrafast molecular nonlinear spectroscopy badly needed a unified approach that could describe and compare the various signals. Numerous confusing terminologies and four-letter acronyms were commonly used to describe the same measurements, and it was not possible at the time to systematically and rationally design an experiment for probing specific processes. My first paper with Yijing Yan and Larry Fried, on applying the Liouville space formalism to the ICN photodissociation experiments of Zewail, was submitted to *J. Chem. Phys*. The referees were very skeptical; one referee said something like "Mukamel has a high reputation, and to save this reputation, he should not publish this useless article." The paper was finally published in *J. Phys. Chem.* in 1989. That paper had started a series of studies that ended up in my book *Principles of Nonlinear Optical Spectroscopy*, which was published in 1995 by Oxford University Press and has helped create a common language that is widely used for the interpretation of signals and for the design of new multidimensional measurements which employ sequences of multiple pulses. These again can be traced back to the stimulating and pioneering work of Zewail.

Author Biography

Shaul Mukamel, a Distinguished Professor of Chemistry and Physics and Astronomy at the University of California, Irvine, received his Ph.D. in 1976 from Tel Aviv University and held faculty positions at Rice University, the Weizmann Institute, and the University of Rochester. His research interests focus on the design of novel ultrafast multidimensional coherent optical spectroscopies for probing and controlling electronic and vibrational molecular dynamics in the condensed phase. He had pioneered the development of coherent multidimensional spectroscopy techniques which span the infrared to the X-ray spectral regimes. His density matrix theoretical framework based on "Liouville space pathways" and his popular textbook *Principles of Nonlinear Optical Spectroscopy* (Oxford University Press, 1995) had created a unified approach for the design and interpretation of these signals. He had employed these techniques to study energy and electron transfer in photosynthetic complexes, excitons in semiconductor nanostructures, and the secondary structure of proteins. Recent extensions include attosecond X-ray spectroscopy, and utilizing the quantum nature of optical fields and photon entanglement to achieve temporal and spectral resolutions not possible with classical light. Nonlinear spectroscopy of molecules dressed by photons in microcavities is studied as well. Mukamel is the author of over 800 publications. He is a Fellow of the American Physical Society and the Optical Society of America and an elected member of the American Academy of Arts & Sciences and the National Academy of Sciences. Recent awards include the Ahmed Zewail ACS Award in Ultrafast Science and Technology (2015), the Coblentz Society ABB Sponsored Bomem–Michelson Award (2016), and the William F. Meggers Award (2017) of OSA.

Chapter 25

The Brilliance of Ahmed Zewail

Paul Midgley*

Ahmed Zewail was a superb scientist, fascinated with the behavior of atoms and molecules, how they reacted, and how best to investigate a world using instrumentation with atomic spatial and femtosecond temporal resolution.

Although I had heard of Ahmed and his work before 1999, it was only after he had won the Nobel Prize for Chemistry that year did I really start to read his papers in detail and to appreciate the groundbreaking work he was undertaking. In the electron microscopy community, Ahmed became much more widely known following his invention of the ultrafast electron microscope (UEM) and when the first papers describing the new instrument and its functionality were published in around 2005. The extension of the ultrafast diffraction method to the electron microscope, incorporating diffraction, imaging, and spectroscopy, was a truly groundbreaking and innovative development, with Ahmed having the foresight and energy to lead the way.

His development of UEM has spawned a whole new era of electron microscopy and, perhaps just as importantly, a whole new generation of electron microscopists who are willing to develop new instrumentation to

---

* pam33@cam.ac.uk

investigate the temporal and spatial dynamics at the atomic and nanoscale. Others in this volume will say much more than I about his corpus of work but, looking back, it is quite remarkable just how productive Ahmed and his group were in UEM in the decade or so after its invention.

Ahmed realized how the UEM could be used to gain new insights into a vast range of physical and chemical behavior using many different modes available in the electron microscope. Just to give a flavor, the UEM has been used to image transient structures in a range of inorganic materials, the dynamics of proteins and macromolecular complexes, magnetic domain nucleation and wall movement using ultrafast Lorentz microscopy, 4D electron tomography and the mechanical vibration of nanostructures, ultrafast scanning electron microscopy, Kikuchi and Bragg diffraction with femtosecond time resolution to investigate lattice vibrations and phonons, the visualization of martensitic phase transformations, and much more. For me, his development of photon-induced near-field electron microscopy (PINEM) and subsequent applications to the study of optical phenomena was especially exciting at the time, as I was also investigating the way electron energy loss could be used to image, in 3D, the optical response of nanoparticles. The remarkable temporal resolution in PINEM opens the door to investigate highly transient phenomena such as the interaction between the electron and the evanescent wave emitted by a nanostructure, for example, after photon excitation. It demonstrated how electron energy gain spectroscopy (EEGS) was viable and how the near-field response of nanoparticle assemblies could be seen with remarkable clarity.

My own research has been inspired by Ahmed's work and philosophy. Although there is no UFM in Cambridge, we do undertake what has been called "multidimensional electron microscopy," or MDEM, combining real and reciprocal space, time and energy space using many different microscopy modes (imaging, diffraction, and spectroscopy) (see Figure 25.1). That multimodal approach, the confluence of spaces, and the rich vein of new information found in the "overlap" region was something of which Ahmed was very much a pioneer.

Unfortunately, I met Ahmed only a few times, but perhaps the most memorable occasion was about ten years ago when I was part of an organizing committee to help celebrate the 75th birthday of Sir John Meurig

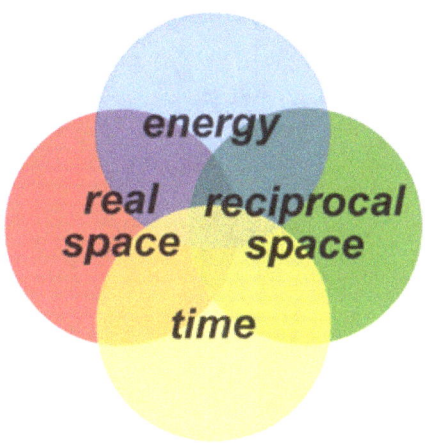

Figure 25.1. The four "spaces" explored by Ahmed Zewail.

Thomas. John and Ahmed were great friends, and John writes elsewhere in this volume about his own memories of Ahmed.

To celebrate John's birthday there was a two-day international symposium, with invited talks by John's scientific friends and colleagues from around the world. Ahmed gave a marvelous opening plenary lecture. The Symposium culminated with a Feast at Peterhouse, the Cambridge College where both John and I are now Fellows, and where John was once Master. In addition to those at the Symposium, some of the partners of the invited speakers were also able to join for dinner. The Feast was something of an event. One of the invited speakers was Prof. Joachim Sauer, whose wife just happened to be the Chancellor of Germany! Ahmed had John on his left and Martin, Lord Rees, Astronomer Royal, and then President of the Royal Society, on his right (see Figure 25.2) and was due to give an after-dinner speech. Shortly before the festivities were due to begin, it was realized that Ahmed would not be able to read his notes for his speech; Peterhouse's dining hall is somewhat dark at the best of times, but especially so on a cold December evening. A rapid search by me of John Lewis department store proved fruitful. A suitable source of photons was found, fortunately not with a femtosecond pulse, but a rather more conventional continuous white light — and Ahmed's speech was illuminated, illuminating and brilliant!

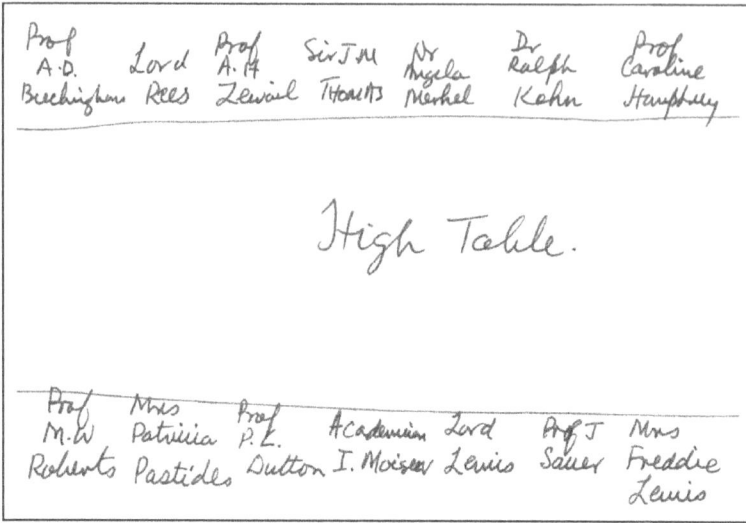

Figure 25.2. Part of the original Peterhouse table plan for the Feast to celebrate Sir John Meurig Thomas' 75th birthday.

Author Biography

Professor Paul Midgley is Professor of Materials Science and Director of the Wolfson Electron Microscope Suite in the Department of Materials Science and Metallurgy at the University of Cambridge. He studied Physics at the H.H. Wills Physics Laboratory at the University of Bristol, receiving his Ph.D. in 1991 for electron microscopy studies of high Tc superconductors. He held two Research Fellowships at Bristol, the first funded by The Royal Commission for The Exhibition of 1851, the second by The Royal Society. He moved to Cambridge in 1997 as an Assistant Director of Research and was then promoted to University Lecturer, University Senior Lecturer, Reader, and finally to Professor of Materials Science in 2008. He is also a Professorial Fellow at Peterhouse. His interests lie primarily in transmission electron microscopy and, in recent years, has focused on the development of electron tomography, precession electron diffraction, and energy loss spectroscopy and their application to nanoscale materials. Recently, he has developed compressed sensing and machine learning algorithms tailored to the analysis of large multidimensional electron microscopy data sets. He sits on the Editorial Boards of a number of journals and was, for many years, Editor-in-Chief of the journal *Ultramicroscopy*. He has received a number of awards and prizes including, in 2004, the Institute of Materials' Rosenhain Medal and, in 2007, the Ernst–Ruska Prize of the German Microscopy Society. From 2008 to 2012, he was President of the European Microscopy Society and, from 2012 to 2016, the Past-President. In 2014, he was elected a Fellow of the Royal Society and, in 2016, an Honorary Fellow of the Royal Microscopical Society.

Chapter 26

Ahmed H. Zewail: Remembering a Hero and Friend in Science

*Jörn Manz**

My first meeting with Ahmed took place in the early 1980s. He had already published his vision of femtosecond chemistry,[1] and he was acquiring a worldwide reputation for excellence in ultrafast laser spectroscopy, which, at that time, meant picosecond spectroscopy. One of the leaders at the time, Wolfgang Kaiser of the Technische Universität München, invited Ahmed to Munich. As a Privat-Dozent in G. Ludwig Hofacker's group of Theoretical Chemistry, I met Ahmed for the first time during that visit. I spontaneously felt admiration for Ahmed as a pioneering hero and also developed a friendship in science which grew over the years. From that time on, I followed his publications with ever-growing fascination. His vision was in wonderful correspondence to the simultaneous change of paradigm in quantum reaction dynamics, inspired by the work of Joshua Jortner and Eric ("Rick") J. Heller, namely from time-independent scattering theory to time-dependent wavepacket dynamics. The revolutionary path into Femtosecond Chemistry during the 1980s was

---

*jmanz@chemie.fu-berlin.de

supported, on the theoretical side, by fundamental developments of the methods that enabled quantum model simulations of molecular processes in real time, and applications to new concepts and phenomena such as the recognition of conical intersections as funnels for femtosecond depletions of electronic excited states, or laser control of chemical reactions.[2] I had the privilege to serve as an active witness of the developments in those exciting days. Finally, when Ahmed published his famous first observation of a chemical reaction passing along the reaction path,[3] I immediately wrote a postcard to him (remember this was the time before emails) saying, "This is the Nobel prize." Ahmed told me later that he kept my postcard on his office table for many years. He also told me that he kept a symbolic painting by my mother in his office, showing sailing boats on the way to new frontiers.

In 1992, after seven years at the University of Würzburg, where I was able to contribute to the developments in ultrafast quantum dynamics, I took the Chair of Theoretical Chemistry in the Freie Universität Berlin. Courageously, but also a bit naively, I asked Ahmed whether he would be willing to present a lecture for the inauguration ceremony. Ahmed said "yes" — but subject to the condition that this should be in the frame of the first international conference on "Femtosecond Chemistry." With enthusiasm and lots of advice from Ahmed and also from Nobel laureate John C. Polanyi, I launched "Femtosecond Chemistry — The Berlin Conference," which took place in March 1993. This was the first meeting in this field that brought together leading scientists from the West and the East after the fall of the Berlin Wall. The proceedings of the conference are documented in the special issue of *J. Phys. Chem.*,[4] co-edited with Welford "Will" Castleman, Jr., and subsequently in the book *Femtosecond Chemistry*, co-edited with my friend Ludger Wöste.[5] Since then, "Femtosecond Chemistry" has been established as a biannual series of conferences, and Ahmed contributed to its success with keynote lectures and strategic advice in Lund (1995), Lausanne (1997), Leuwen (1999), Toledo (2001), Paris (2003), Washington, D.C. (2005), Oxford (2007), Madrid (2011), and Hamburg (2015), with two exceptions, namely Beijing (2009), because he had to give priority to his duties as advisor on international relations in science to US President Barack Obama, and Copenhagen (2013), when he was already sick.

Ahmed had a profound impact on the development of quantum reaction dynamics in Germany that is little known. In 1994, my colleague Jürgen Korsch (Universität Kaiserslautern) and I applied as coordinators for support from the Deutsche Forschungsgemeinschaft (DFG) for an interdisciplinary network (Schwerpunkt-programm) of about 30 groups from Theoretical Physics and Theoretical Chemistry, entitled "Time-dependent quantum phenomena and methods in physics and chemistry." It so happened that the responsible authorities of the DFG met with Ahmed to ask his advice, and he supported our project enthusiastically. This added coherently to the positive evaluations by the German referees. In the end, the project was supported for seven years and produced immense scientific output. For example, ten of the young investigators of that network were promoted to Professorships in Theoretical Physics or Theoretical Chemistry, over the lifetime of the project. Today I would say proudly that it strengthened Germany's position as one of the leading countries in the theory of quantum reaction dynamics at that time, including the groups of W. Becker (Berlin), V. Bonačić-Koutecký (Berlin), L.S. Cederbaum, together with H. Köppel and H. D. Meyer (Heidelberg), R. de Vivie-Riedle (Berlin, now Munich), W. Domcke (Munich), V. Engel (Würzburg), E.K.U. Gross (Würzburg, now Halle), P. Hänggi (Augsburg), B. Hartke (Stuttgart, now Kiel), M. Holthaus (Marburg, now Oldenburg), R. Jaquet (Siegen), O. Kühn (Berlin, now Rostock), U. Manthe (Freiburg, now Bielefeld), P. Saalfrank (Potsdam), R. Schinke (Göttingen), W. Schleich (Ulm), B. Schmidt (Berlin), R. Schmidt (Dresden), M. Schreiber (Chemnitz), and G. Stock (Frankfurt, now Freiburg). Thank you, Ahmed!

The most heartening event for me on Ahmed's way to the Nobel Prize for Femtosecond Chemistry was the Nobel Symposium in Björkborn, Sweden, in 1996. I had the honor of presenting the third lecture, after Ahmed's and Nobel laureate R. A. Marcus', with an overview of the development of quantum wavepacket dynamics simulations of molecular processes and chemical reactions, from the day of Schrödinger (1925), who suggested the fundamental Schrödinger equation,[6] to the subsequent implementations and applications over seven decades. To prepare the lecture, I invested about nine months, consulting more than 50 colleagues worldwide, with the great support of Dr. Werner Gans, who visited a dozen scientific libraries in Berlin to dig out the visionary early contributions. It was my pleasure to make clear that Ahmed's observations of chemical reactions with femtosecond time resolution provided the experimental breakthrough in this field that stimulated myriad complementary quantum dynamical simulations. My lecture was published[2] with 1,500 references. After the Nobel committee called Ahmed in the fall of 1999 to inform him that he was the recipient of the 1999 Nobel Prize in Chemistry, in recognition of his discovery and development of Femtosecond Chemistry,[7] there must have been at least two days of celebration of the achievements of Ahmed and his team and partners at Caltech in Pasadena, and many, many congratulations from all over the world. After those two days, I was truly moved when Ahmed called me, saying that he was of course very happy but now also very, very tired, but before going to bed he wanted to say "thank you, too" to me.

I close these reminiscences with an anecdote that Ahmed told me concerning his relations with Germany, about which one may smile, or frown: As a pupil in Egypt, Ahmed was an enthusiastic scholar of the German language and participated in a language contest that offered a visit to Germany for the winner. The final hurdle took place in the German Embassy in Cairo. The German Ambassador handed Ahmed a book and asked him to read from it aloud. After a couple of minutes, the Ambassador said *danke* (meaning "thank you"). Now the German word *danke* has different meanings, depending on the situation. Ahmed interpreted it to be the Ambassador's expression of approval of his reading, so he went on reading. After a while, the Ambassador said again *danke*, now in a demanding tone. Ahmed thought that the Ambassador was pleased even more, and went on reading. Finally, the Ambassador stopped him, and eliminated him from the list of the candidates, because in his evaluation, Ahmed did not even know the German word *danke*. I feel ashamed for that Ambassador of my country, who did not recognize the potential of the highly gifted pupil, and who did not consider the possibility that the word *danke* could be misunderstood. Well, this event had many consequences. Of course, Ahmed did not visit Germany. Instead, he went to the US, and eventually to Caltech, where he stayed, working for his Nobel Prize as the second scientist of Islamic origin to collect that high honor. Later, after he received the Prize, Ahmed was offered the position of director of Max Planck Institut für Biophysikalische Chemie in Göttingen, which had previously been headed by Nobel laureate Manfred Eigen. Well, the Ambassador's high-brow attitude may have had its long-time psychological effect that contributed to Ahmed's declining the offer and remaining at Caltech.

Today, I would like to join all scientists in Ahmed's field, Femtosecond Chemistry, saying with admiration: "*salam*, Ahmed, hero and friend, and truly '*danke*,' from the bottom of my heart!"

References

1. A.H. Zewail, "Laser Selective Chemistry — Is It Possible?" *Phys. Today* **33**, 2–8 (1980).
2. J. Manz, "Molecular Wavepacket Dynamics: Theory for Experiments 1926–1996," in *Femtochemistry and Femtobiology: Ultrafast Reaction Dynamics at Atomic Scale Resolution*, Nobel Symposium, Vol. 101, V. Sundström (Ed.) (Imperial College Press, London, 1997), pp. 80–318.
3. M. Dantus, M. J. Rosker and A.H. Zewail, "Real-time Femtosecond Probing of 'Transition States' in Chemical Reactions," *J. Chem. Phys.* **87**, 2395–2397 (1987).
4. J. Manz and A.W. Castleman, Jr. (Eds.), Special Issue "Femtochemistry," *J. Phys. Chem.* **97**, 12423–12644 (1993).
5. J. Manz and L. Wöste (Eds.), *Femtosecond Chemistry*, Vols. 1 & 2 (Wiley-Verlag Chemie, Weinheim, 1995).
6. E. Schrödinger, "Quantisierung als Eigenwertproblem," *Ann. Phys. (Leipzig, Ger.)* **81**, 109–139 (1926).
7. A.H. Zewail, "Femtochemistry: Atomic-scale Dynamics of the Chemical Bond Using Ultrafast Lasers (Nobel Lecture)," *Angew. Chem. Int. Ed.* **39**, 2586–2631 (2000).

Author Biography

Jörn Manz (b. 1947 in Hamburg, married, three children and two grand children) studied physics at the universities of Hamburg and München (1966–1970), and then he switched to Theoretical Chemistry where he did his Ph.D. with G.L. Hofacker at Technische Universität München (TUM, 1972). After two postdoctoral periods with R.D. Levine (Weizmann Institute, Rehovot, 1974–1975) and with D.J. Diestler (when he was Alexander von Humboldt awardee at TUM, visiting from Purdue University, 1975–1976), he did his Habilitation at TUM (1978). In 1985, he was saved for our science by the German Fiebiger emergency program for young scientists, which catalyzed his call to Universität Würzburg as C3-Professor for Theoretical Chemistry. From 1992 till 2012, he was full (C4-) Professor of Theoretical Chemistry at Freie Universität Berlin (FUB). After retirement from FUB, he became senior research guest Professor at Shanxi University in Taiyuan (four months p.a.) while also continuing research as guest Professor in the group of his successor, Professor Beate Paulus, at FUB (eight months p.a.). His scientific interest is the Quantum Theory of Molecular Reaction Dynamics, including Femtosecond and Attosecond Chemistry, in pleasant and fruitful cooperation with brilliant group members and excellent international partners. In 1993, he organized the first international conference on "Femtosecond Chemistry" which since then has been established as a bi-annual series of conferences. From 1995 till 2001, he was coordinator of the interdisciplinary German network on "Time dependent quantum phenomena and methods in physics and chemistry," including about 30 groups, together with H.J. Korsch (Theoretical Physics, Universität Kaiserslautern.) He was founding member of the Berlin network (Sonderforschungsbereich), including about 15 groups supported by Deutsche Forschungsgemeinschaft (DFG) on "Analysis and Control of Ultrafast Photo-Induced Reactions" (1998–2010). From 2005 till 2013, he coordinated the trilateral project on chemical reactions via conical intersections, for cooperation with scientists from Israel and Palestine, supported by DFG. From 1993 till 1999, he also served as Director of the Institut für Physikalische und Theoretische Chemie at FUB, and from 2000 till 2013 as Professor of liaison (Vertrauensdozent) for highly talented students supported by Studienstiftung des Deutschen Volkes. He has profited a lot from, and acknowledges with gratitude, the scholarships from Studienstiftung des Deutschen Volkes, Minerva Stiftung, Chemie Dozenten-Stipendium des Verbandes der Chemischen Industrie, and additional financial support, in particular from

DFG, Fonds der Chemischen Industrie, Stiftung Volkswagenwerk, scientific programs of the European Community, the Japan Society for the Promotion of Science (JSPS), and Deutscher Akademischer Auslandsdienst (DAAD). In 2009, he was elected as member of the German National Academy of Sciences "Leopoldina."

Chapter 27
Ahmed Zewail: A Reminiscence

*Paul E. Dimotakis**

Some years ago, sitting at one of the Caltech Athenaeum faculty round tables, a colleague lamented that the days of new Caltech Nobel prizes were over. I disagreed and when challenged, I mentioned that I would not be surprised if Ed Lewis and Ahmed Zewail, for example, would likely be so honored. They were, in fact, in 1995 and 1999, respectively.

Nowhere is it written that people with unusual intellectual and scientific ability possess superior qualities in other dimensions. However, both Ed Lewis and Ahmed Zewail left behind a better world because they were also outstanding in their character and human qualities. Ahmed was remarkable in many ways but I will leave it to others better qualified to comment on his scientific achievements.

At the Athenaeum table, he contributed knowledge and wisdom, especially about the Middle East. He deeply felt the need to help in Egypt, participating in the formation of new science initiatives and institutions there. During and following the Arab Spring events, he offered his time, energy, and wisdom there to help as best he could. While overtaken by circumstances beyond his control that thwarted those initiatives, they made a difference, perhaps planted seeds for new beginnings later, and

---
*dimotakis@caltech.edu

mark his sense of responsibility to his country of origin. He had a keen sense of priorities; it impressed me that he basically refused to use email, which he considered as imposing a waste of precious time.

In other contexts, I served with him in Caltech committees, in which he played a pivotal role and was responsible for leading the formulation of the Institute's then policy on large efforts/centers that was adopted.

We were overjoyed when it appeared he might overcome the health challenges that arose and felt like he and we were given a second chance and opportunity.

We miss him.

Author Biography

Paul Dimotakis came to Caltech from Greece, with family origins in Crete, an island closer to Ahmed's country of origin than to mainland Greece. He received a B.Sc. in Physics, an M.Sc. in Nuclear Engineering, and a Ph.D. in Applied Physics, all from Caltech. He joined the Caltech faculty and is presently the John K. Northrop Professor of Aeronautics, Professor of Applied Physics, and Senior Research Associate of the Jet Propulsion Laboratory, at the California Institute of Technology. He and his collaborators have contributed to fluid dynamics and turbulence, combustion, instrumentation and optical diagnostics, greenhouse-gas monitoring, and other topics. He is a consultant to the Aerospace industry, contributed to the sail design of America[3] in their successful defense of the Americas Cup, and leads or contributes to studies for the US government. He is a Fellow of the APS, the AIAA, and the AAAS, and a member of the National Academy of Engineering.

Chapter 28

"Stop All the Clocks"

*Dmitry Shorokhov**

An Attempted Apology

Sad but true: we never know what people actually mean to us until we finally lose them. It goes without saying that human nature suffers from numerous imperfections, such as selfishness, impatience, and short-sightedness. As another example, we often tend to take things around us, including generosity of our parents, spouses, and mentors, more or less for granted, assuming blindly that this is what we probably deserve. In reality, however, regardless of whether or not we do indeed deserve having what we have, those very special ones who *care* lend their helping hands to us without drawing a balance sheet — even at times when we are not acting virtuously enough and/or hurt their feelings. That may sound pretty trivial, but we often tend to forget this — or, without generalizing too much, it is probably fair to say that I do, as I keep learning this lesson again and again in life. Another important thing I have realized as I have been growing older is that our attention span is often too limited for the true feeling of empathy to develop: we rush through conversations and skip over what we consider to be "irrelevant details" in an attempt to substitute a simplistic model of some sort for the real thing. I wish I could apologize to

---

*dikun@caltech.edu

Prof. Zewail for all of the above, but it is, apparently, too late to do so now. The only source of comfort I am left with in that regard is the statement he used to make whenever he felt upset with what was going on around him: "one cannot expect receiving too much gratitude from one's own children."

Of Prof. Zewail's numerous colleagues and mentees I was, arguably, the closest one to him, interacting with him daily on a one-on-one basis and assisting him on a personal level in his workplace. A substantial fraction of my job as a senior scientist here at Caltech was to prepare his talks for presentation and books for publication, and to maintain the semi-professional IT infrastructure, including the (virtual) private network environment and sophisticated network-attached storage servers, that I had created on his request. Because, due to my skewed scope of interests and somewhat unusual educational background, I did happen to possess the technical expertise required to do the above things quickly and easily, I did feel an urge and — I do not use the word lightly here — a moral obligation to help him, but I often regretted having to spend time on things that did not have much to do with fundamental science as I understood it (i.e., "building insights and breaking boundaries"). Although the answer had always been pretty obvious, I often used to ask myself: "why me again?" without fully realizing, until very recently, that paying that little for being able to operate right next to him as if I were an independent PI — and to pursue my own research agenda at Caltech for more than a decade under his umbrella — was a truly unique, *unprecedented* luxury.

Now that his office upstairs is empty, it feels like half of my universe has collapsed.

Meeting the Protagonist

At Caltech, Prof. Zewail has long served as the (first) Linus Pauling Chair Professor of Chemistry and Professor of Physics. With generous support of the Gordon and Betty Moore Foundation, he has been closely supervising, and directing, the cutting-edge experimental and theoretical research performed at the Physical Biology Center for Ultrafast Science and Technology (UST). In 1999, he was proclaimed the sole recipient of the

Nobel Prize in Chemistry with the following citation: "for his studies of the transition states of chemical reactions using femtosecond spectroscopy." During the past decade or so, instead of resting on his laurels, Prof. Zewail went on to develop the completely novel field of 4D ultrafast single-electron microscopy imaging (4D UEM) that focuses on direct visualization of structural dynamics and chemical kinetics taking place in the four dimensions of space and time. Such 4D visualization of the initial steps of physicochemical transformations of matter with the femto-to-attosecond temporal resolution is of paramount significance in elucidating the fate of crystal, amorphous, and (macro)molecular structures as they undergo phase transitions and/or chemical reactions in real time. The field is now well established, with laboratories around the world, including those in Minnesota, Purdue, Harbin, Osaka, Giza, Lausanne, Lansing, Göttingen, Ulsan, and Stockholm, among others, using 4D UEM as the method of choice in their studies.

Despite his world fame and enormous achievements extending well beyond a typical human scale, Prof. Zewail was truly concerned with the well-being of his students, employees, and colleagues, as well as with that of the entire science community and — last but not least — developed *and* developing nations at large. The lucky few who have been chosen to assist him in pursuing his goals will always treasure the memories of his willingness to listen (and to understand), his truly unique sense of humor, and the remarkable warmth of his personality. What made him so special as a group leader and research advisor was, in my opinion, his capability to meticulously orchestrate the work of a vast, aspiring, and ambitious team by making everyone of its members do what they were really good at — and what they enjoyed doing *the most*. Also, as a person driven by sheer curiosity (hence his deeply routed and genuine respect for those who devoted their lives to the quest for fundamental knowledge), he would not hesitate to let his coworkers try a number of things prior to presenting him with real, significant accomplishments. "Why argue with me if both of us know that I am right?" was, no doubt, one of his killing arguments, but, like a wise politician reluctant to employ brutal force, he would consider imposing his will on people to be a last-ditch measure and he did not at all enjoy doing so.

The Bigger Picture

What makes America strong is its enormous innovative potential originating from the open-mindedness and cultural diversity characteristic of its inhomogeneous society. Indeed, there is a lot more to our world than uniformity, orthogonality, and rectilinearity — suffice it to take a look at the Google maps of Paris, Augsburg, or Tokyo. As a true citizen of this *curvilinear* world spanning both Western and Eastern hemispheres, Prof. Zewail had no problem thinking outside the boxy box, what used to be one of his most significant power points, another one being his humble attitude toward life as a whole, and toward processes of acquiring new knowledge of fundamental significance pertinent to it in particular. Thus, he would passionately argue that, because the human brain is very limited in its capacity to represent and process things extending beyond what we are exposed to in our daily lives, there could well exist phenomena *incomprehensible* to us as scientists. The remarkable efficacy of his cognitive approach was stemming from prioritizing concepts over details. On yet another note, he was truly fond of life in all of its manifestations, and he was really grateful for the gift of it — hence his jovial curiosity, resilience, and remarkable optimism. The above features constituted the foundation of a strong human touch implicit in Prof. Zewail's personality, and they were, of course, to a very significant extent shaping up his life and his startling career — as world-renowned scientist and widely recognized statesman alike.

With that in mind, it would perhaps be instructive to reflect on how Prof. Zewail's research group was being run, and what essential factors contributed to its unparalleled efficiency and unprecedented productivity. The number one thing to bear in mind in that regard was his direct involvement in all the group affairs — be it scientific research, personal matters, or interconnections within, or between, the individual subgroups. Sadly, there exist numerous labs where researchers need to make arrangements for half-an-hour-long meetings with their immediate advisors weeks in advance, let alone the fact that there are a multitude of topics that cannot be addressed during such meetings. There are laboratories — both inside and outside the US — where postdoctoral scholars are supposed to raise money, and within such labs there may exist areas of research dominated by certain "established" individuals. Remarkably, none of the above was an issue at the UST as it resembled an extensive family encompassing the father surrounded by his sons and daughters (an analogy often used by Prof. Zewail himself). Money, equipment, and laboratory space were in abundance and, last but not least, everyone was treated on an equal footing. The only price to pay for all of that glorious privilege was being 100% loyal toward the UST — being, e.g., 99% loyal would not suffice.

The 100%-loyalty policy implied, among other things, that Prof. Zewail was the only one to make important decisions, and that the final word was his in all the group affairs that were catching his eye — or that were brought to his attention otherwise. As an example, determining what the sequence of authors' names on a research paper was going to be like had always been Prof. Zewail's prerogative (upon completion of a project, he would meet all of its contributors, let them know his decision, and ask them whether or not there were any objections). It is my understanding that, in a lab comprising world-class researchers striving to achieve, preventing people from putting their names on the articles directly was the only way to avoid creating severe conflicts of interest. Importantly, Prof. Zewail would typically try to ensure that every one of his postdocs would have a fair chance to publish at least one paper with him as a coauthor. Therefore, publishing a research account after a year or so spent at the UST was a relatively easy thing to do, but the subsequent contribution implying that its authors were there to stay could follow years later, and getting it out would typically represent a challenging task. Given that the overall output of the Center would not exceed ~20 articles per annum, securing the annual slot for oneself could easily turn into a highly nontrivial quest, but people did not seem to mind that at all.

Finally, a word is appropriate here concerning the role of "outreach" in raising awareness about the research activities performed at the UST. Despite the busy schedule, Prof. Zewail paid a great deal of attention to popularization of his ideas, not only because receiving a Nobel Prize implies a certain degree of publicity but also because dissemination of fundamental knowledge is known to prevent mediocritization, complacency, and stagnation from spreading. He was extremely concerned with the quality of the work, and he would typically ask us to double-check and triple-check things if he happened to have a slightest doubt about it, but research presentation quality was almost equally important to him. Although coming up with an "eye-catcher" has always been one of his major objectives, he would never go over the border when advertising his findings. Quite the contrary, the reason why he insisted that certain people redraw their figures over and over again was, mainly, the lack of clarity, and not necessarily the poor graphics. The same was true for manuscripts as well: it would often take him a couple of dozen editing rounds to prepare a paper for publication. During our one-on-one conversations, he was bitterly complaining that most people do not know — and do not really care to understand — how to make their write-up clear, concise, logical, and appealing to the reader. My own attitude was quite simple: "either you have it, or you do not," but he kept insisting that presenting one's results the appropriate way was a critical skill every senior person was supposed to master: "one cannot afford presenting a one-hundred-thousand-dollar project the way a ten-thousand-dollar project is typically presented to the public."

Zooming in: "On a Personal Note"

"No man is an island, but some of us are long peninsulas." Due to my personality type, I have never been connected to too many people, and Prof. Zewail was willing to respect that. Moreover, because I needed to be alone "to decompress and process," he was the one who made me occupy an isolated, spacious office of my own, and putting a wall sign reading "THINK—SILENCE PLEASE" (in capital letters) over its door was, actually, his idea. Following my relocation, he did mention to me that keeping the door of the office ajar rather than closed could be a nice thing to do — and that doing so would, in fact, be in my best interest — but when I informed him that I was going to keep it locked at all times, he did not seem to have any major objections. A year later, following my promotion to a senior position within our Center, Prof. Zewail attempted to make me take over the gas electron diffraction subgroup of the UST, but I had a strong feeling that the experimental research that was being performed there was a dead end, and I already had my own agenda associated with molecular biology by that time. The line of research I had been pursuing was purely theoretical — which was a big disadvantage because Prof. Zewail's heart has always been with experiment, not with theory — but I grew very attached to it, and I was not at all willing to give it up. We had an emotional conversation during which I happened to ask the following rhetorical question: "If I am not allowed to pursue my own research, then what am I doing here?" Because what I did during that meeting was a "no, no" kind of thing and I was, of course, perfectly aware of it, I was prepared to leave, but, remarkably, Prof. Zewail decided to give me a chance. We never talked about that conversation again — ever — but it is my sincere hope that our joint accomplishments spoke for themselves.

It is always fun to watch things migrating from the science fiction world into our daily reality. As a long-time leader of the UST theory subgroup, I have been utilizing a supercomputer cluster to follow structural transformations of biological macromolecules for about a decade now. With hundreds of independent transformation trajectories generated during the course of massively distributed, all-atom molecular dynamics (MD) calculations, we are currently capable of performing ensemble-convergent numerical simulations of systems possessing numerous degrees of mechanical freedom — and immersed in aqueous environments — with an unprecedented atomic-scale spatiotemporal resolution. Following careful testing of our findings against ultrafast T-jump spectroscopy results, Prof. Zewail named the approach *4D computational microscopy* because the statistical certainty and signal-to-noise ratio characteristic of our data were comparable to those typically obtained in laboratory experiments. The statistical convergence mentioned above is an essential feature of such numerical experiments because, when a multitude of transformation trajectories are calculated under identical ambient conditions, random fluctuations average out and ensemble-wide tendencies characteristic of biologically relevant behaviors emerge. A variety of sophisticated theoretical models and coarse-graining techniques pioneered by members of the UST theory subgroup are currently being exploited in our research to pinpoint, visualize, and explain the biological phenomena observed "in silico."

A couple of decades ago, Prof. Zewail entered the uncharted waters of molecular biology in an attempt to discover the driving mechanisms that underlie its overwhelming complexity — the mechanisms that, despite all the clumsiness of biological macromolecules, make these amazing beasts cooperative and robust. As was pointed out by Feynman in one of his essays, there had always been "something about biology: it was very easy to find a question that was very interesting, and that nobody knew the answer to." "Whatever you dream, begin it, for boldness has genius, power, and magic in it" was one of Prof. Zewail's numerous mottos. Given his remarkable perseverance, it is, perhaps, not at all surprising that he has been able to identify, and address, a number of biological questions of fundamental significance, but what I am accentuating below is his perception of what exactly had to be figured out. Putting a number of our conversations into perspective, I recall him repeatedly stating that (i) very complex mechanisms cannot be very efficient by definition, and (ii) if one is incapable of describing a phenomenon in simple terms, one does not really understand much about it — period. To give us a flavor of what he was eager to see, he would often use the following analogy. Imagine an $n \times n$ correlation matrix $\mathbf{X}$, $x_{i,j} \neq 0 \ \forall \ i, j \leq n$, representing a tightly knit biological system. Depending on the nature of the system, one may be able to block-diagonalizable $\mathbf{X}$ by disregarding its less-significant terms (which is often the case). In a number of our joint biological studies, we have been able to push the point made above a bit further by reducing the entire N-dimensional configurational space, $N \to \infty$, to just one or two variables *comprehensively* parameterizing the behavior of interest.

Epilogue

Prior to leaving Caltech at the very beginning of July 2016, Prof. Zewail mentioned to me, among several other things, that he had had an intention to prepare a book for publication with me as a coauthor. The book was supposed to have a biological twist, and, because I felt very strongly that it was a timely and exciting idea, I was really looking forward to his return. The startling news announced a month later left me completely devastated as I had great difficulty believing that a man so full of energy, life, and creative ideas could possibly pass away so suddenly and so tragically soon. Several months have elapsed since Prof. Zewail's passing, but I must admit that I am still struggling to comprehend the loss. Nevertheless,

I am grateful to the Editors of the present volume for insisting on its prompt publication because, no matter how painful it may be for many of us to go through our reminiscences right now, it needs to be done as quickly as possible in order to preserve precious memories from fading. Because Prof. Zewail was like an immediate family member to me, I have a lot to tell. I could go on and on, sharing the memories of how he rushed me, then a first-year postdoc, to a hospital in his fancy sports car, describing in detail how he unexpectedly visited a house concert organized by my wife at our apartment in Pasadena, or reflecting on my numerous arguments and conversations with him here at Caltech, but the amount of space available to me is subject to a limit.

In conclusion, I would like to stress that outstanding people are somewhat similar to large planets as they are, by their sole presence, capable of inducing highly nonlinear space–time effects. Throughout the above write-up, I have taken every effort to be fair and unbiased, but, as one of Prof. Zewail's closest satellites, I cannot help but realize that I still tend to view the world around me through his eyes.

Acknowledgments

I am indebted to Dr. Dema Faham and Dr. Milo Lin for their helpful comments.

Author Biography

Dmitry Shorokhov was born in Russia in 1971. He received his B.S. (1992) and M.S. (1995) in Applied Mathematics and Physics from the Moscow Institute of Physics and Technology (the renowned "Phystech"), arguably the number one technical school in the Soviet Union that had been founded back in 1946 with the purpose of educating key research personnel for the projects of "utmost importance," many of them classified. Thanks to unmatched flexibility of so-called "Phystech System" largely modeled on Caltech curriculum as originally proposed by Phystech's Founding Fathers (Kapitsa, Landau, and Semyonov), he was allowed to present a Master's degree thesis in Statistical Mechanics while majoring in Electrical Engineering and Computer Science. Following a one-year-long practicing stint in gas electron diffraction in his home town of Ivanovo (1995–1996), he had been admitted to graduate studies at the University of Oslo in Norway, where for his Ph.D. in Physical Chemistry (2000), he studied gas-phase structures of small, conformationally flexible molecules, including a DNA base. Upon receiving a postdoctoral research fellowship from the Alexander von Humboldt Foundation, he spent two years at the Technical University of Munich, followed by another two years at the University of Augsburg in Germany, performing experimental and theoretical electron density distribution studies in the gas phase and in the solid state. Another spectacular turn in his career came about in 2004, when Prof. Zewail invited him to join the research team of the Laboratory for Molecular Sciences in Pasadena as a postdoctoral scholar. While at Caltech, he has been overseeing the acquisition, construction, and operation of a massively distributed computing facility as a senior postdoctoral scholar (2007), staff scientist (2007), and senior staff scientist (2008). At present, his research interests involve experimental and computational molecular biology as well as high-performance computing.

Chapter 29

My Time with a Giant

*Dongping Zhong**

On August 3, 2016, I arrived in Singapore to arrange a conference in celebration of Ahmed H. Zewail's 70th birthday at Nanyang Technological University. When I turned on my phone, I was shocked to see hundreds of e-mails from friends all over the world. I immediately realized something serious had occurred. When I opened the messages, my body went numb, my hands shook, and my eyes filled with tears. My friends informed me that Professor Ahmed H. Zewail had just passed away in Pasadena (on August 2 in US). I was shocked. As my emotions overcame me, I reflected upon not only the momentous contributions Zewail has made to the scientific community, but also the lasting impact he had on our lives.

The next day, I posted several pictures of Ahmed Zewail, some with me, on my WeChat and wrote, "Mourning my great mentor Ahmed H. Zewail: A visionary leader, a brilliant scholar, a giant scientist, a deep thinker and a faithful son of Egypt." After searching for all possible flights, I could not make it to the memorial service in Pasadena on August 4, and instead my wife, Lijuan Wang, immediately went in my place. On that day, a giant fell. Science lost a great leader and we lost our dear friend. Dr. Ahmed Zewail left us too early.

---

*dongping@physics.osu.edu

I first came across Dr. Ahmed Zewail's work in his 1988 review article published in *Chemical & Engineering News* with the late Richard Bernstein. This article was faxed to my former group in Fudan University, where I was a Ph.D. candidate studying laser chemistry. Due to the significant delay of receipt of journals to China at the time, we had no way of knowing Dr. Zewail's revolutionary work in time, such as the work published in the *Journal of Chemical Physics* about his first groundbreaking experiment on clocking of chemical-bond cleavage (I–CN). My advisor's wife, visiting Berkeley then, heard the story and sent the review to us immediately. Because I studied the colliding pulse mode-locked (CPM) ring dye laser during my Master's period, I was asked to give a talk in the group about the review, even though I was full of questions. I was superexcited by the new breakthroughs and brave enough to write a letter directly to Dr. Zewail and expressed my desire to join his group in the near future. After several weeks, I was thrilled that I received a reply letter from him encouraging me to apply for the chemistry graduate program in Caltech. Of course, at the time, we did not have e-mail or internet, so our correspondence was through traditional letters. You can imagine the excitement a young graduate student in China during the late 1980s experienced from receiving a letter from Professor Ahmed Zewail at Caltech! However, for certain reasons in China in 1989, I went to Kansas State University to work with Professor Donald W. Setser on chemical reaction dynamics. I promised myself to work there for at least three years. In early 1993, I wrote to Dr. Zewail again and told him that I had read all his femtochemistry papers. I asked many questions about the papers and then expressed my excitement and wish to join his group. I believe that Dr. Zewail got excited too and I transferred to Caltech (he told many times later how impressed he was by my studying of all his papers)!

I arrived at Caltech in late December 1993 and ran into a big earthquake. I was completely amazed by the femtosecond setup; the experiment was still working in the sub-basement while the floating optical table swung due to the quakes. I was working with a postdoc, Dr. James Cheng, to develop the iodine-atom detection in a supersonic molecular beam with femtosecond resolution. With painful dye lasers, we needed to maintain the laser pulse not only with enough femtosecond resolution but also with enough narrow bandwidth to resonantly detect iodine atoms. We finally succeeded after several months and developed a mass spectrometry with four resolutions of time, energy, direction, and state. Dr. Zewail was extremely happy and asked us to reexamine the dissociation reactions of iodine-containing molecules (I–CN and I–Hg–I) that had been studied in the group before by the laser-induced fluorescence method. We further studied a series of new molecules (CH_3–I, C_6H_5–I, and I–C_2F_4–I, C_6H_5–I…Cl_2) with different reactions. When we told him that we were able to study bimolecular reactions of iodine molecule (I-I) with a series of electron-rich organic molecules, he got very excited. We directly pumped the bimolecular charge-transfer band and launched the bimolecular reaction exactly into the transition state and followed the entire evolution of the reaction by monitoring the product of iodine atoms. The classic system is the reaction of benzene with iodine molecule (C_6H_6…I_2), also later highlighted in the Nobel press release. During that time, Dr. Zewail was filled with anticipation. In his excitement, he called the laboratory from a conference in Paris, and on another day, from his office at 5 AM. When I graduated in 1999, Dr. Zewail was very satisfied with my work and informed me that he would recommend me for the Herbert Newby McCoy Award in the Division of Chemistry and Chemical Engineering and the Milton and Francis Clauser Doctoral Prize in Caltech. Both awards were for most outstanding and original thesis. I was thrilled to receive the highest honor for graduating students at Caltech (Figure 29.1).

My time as a graduate was intense, exciting, and rewarding, akin to the metamorphosis of a cocoon transforming into a butterfly. In less than six years, I learnt everything from Dr. Zewail, from the critical thinking of a project, to simplifying the system's complexity, to crystallizing a new concept or a simple picture. We worked very hard and often drew figures on a restaurant table on the weekend or discussed the manuscript at midnight in his house when he was alone during the summer. We often finished a paper in a few days, from discussion of drawing figures to submission of the manuscript. One of the most important things he emphasized was to summarize our findings in one or two sentences when we finished a new paper. He often told us if you could not explain in layman's terms, then that meant you did not understand the system. As we all know, he had an incredible gift to distill and explain a complex system into a simple idea. Once, Dr. Zewail asked us what makes a great scientist.

Figure 29.1. Taken on the author's graduation day on June 11, 1999, after the commencement.

The answer is that a great scientist has a vision. He had a great vision and led us to a blooming golden age of femtochemistry in the 1990s.

When I graduated in 1999, my interest moved to femtobiology after extensive studies of elementary chemical reactions in real time. I was interested in various ultrafast dynamics in biology, especially protein dynamics and enzyme catalysis. Dr. Zewail had just established a Laboratory for Molecular Sciences funded by the National Science Foundation on studies of fundamental processes in complex systems.

He initially asked me to continue in his group and perform ultrafast electron diffraction. I, however, was attracted to ultrafast biological dynamics and convinced him to let me pursue this field. I finally had complete freedom. On one early morning in October, I was doing experiments in the laboratory and suddenly heard the news that Dr. Zewail had received the 1999 Nobel Prize in Chemistry. I rushed out of the lab and shouted in the sub-basement but there was no response! When I came back after my breakfast, Caltech had just started a press conference in the faculty club Athenaeum. I saw the excitement in many people's eyes. In the coming two months, I was lucky enough to have constant interactions with Dr. Zewail to prepare all of his figures for his Nobel lecture with the multimedia office in Caltech. For each figure, he discussed with me the content, size, color, legend, and overall layout. As we all know, Dr. Zewail had an extremely high standard and was a perfectionist. The Nobel lecture is a masterpiece, highlighting all critical concepts and achievements of femtochemistry accomplished at Caltech with his more than 20 years of enduring efforts. When he returned from Stockholm, we celebrated this great recognition at Athenaeum in January 2000, a memory I will cherish for life (Figure 29.2). In the following three years, he was very busy with all kinds of invitations, and I became more independent in my research. During this time, I, with my labmate, Dr. Samir Pal, started to study biological water using natural amino acid tryptophan as an optical probe, and Ahmed was very excited. The topic continued, and we both had longtime collaborations even after I became independent. I was also fascinated by another important molecule of flavin in biology that led to my own future extensive work on flavoproteins. During that time, we often wrote our papers together at the conference or on the weekend. In such a short time, we published six PNAS papers on protein dynamics. Over the eight years I was in Ahmed's group, we wrote 18 papers together. After his Nobel Prize, he was thinking much faster and deeper. Many ideas naturally flowed out of his mind, and we often finished our manuscripts very quickly and efficiently. He had become a master with a great mind. The three-year postdoctoral training was critical for my transition from a hard-working graduate student to an independent researcher, ready for my own career.

In strong support of my faculty search, he encouraged me to consider multiple departments, including chemistry, physics, and even medical

Figure 29.2. At the celebration party of his Nobel Prize at Caltech's Athenaeum in January 2000 when Ahmed returned from Stockholm.

school. I received more than 20 interviews and finally finished 16. When I was ready to leave for my new position at The Ohio State University in the summer of 2002, he invited my wife and me to his house for dinner (Figure 29.3). I avidly sought his advice for my future, especially during the tenure-track period. Today, I still vividly remember what he told me: (1) Do not be involved in any politics in the department and be focused; (2) Be careful to collaborate with other faculty and be independent; and (3) Do not get stuck on one project for too long and get published quickly. He was absolutely right and I think his advice is invaluable to all assistant professors at any university. I could tell that Ahmed was very confident about me for my new position and we shared many common things: loyalty to our homeland and friends, dedication and focus, and an emphasis on hardwork.

From 2002, we always met each other at least one or two times a year, either by my direct visit to Caltech or at various conferences such as the American Chemical Society annual meeting or the Femtochemistry biannual meeting. We continued to collaborate on biological water and published another three papers on this topic. His brilliant vision, insights, and clarity on the complexity of hydration dynamics have reshaped our

Figure 29.3. At Zewail's backyard in August 2002, when the author was leaving Caltech.

thinking of the protein hydration field. I still remember clearly that we spent an entire day in his office discussing our 2006 PNAS paper sentence-by-sentence. Many critical thoughts in that paper laid the foundation for future work. He was always excited about our research and often sent the e-mails on our published work. I also followed all his great work and was extremely excited about his major breakthrough on 4D electron microscopy in 2005 with brilliant design to combine temporal and spatial resolutions together. He completely opened another new field and we almost witnessed his another walk to Stockholm. He was so close!

Ahmed and Dema made two visits to China in 1996 and 2004 through my initiation. After they attended Yuan Lee's 60th birthday in Taiwan, they went to Beijing for their first visit to China in 1996. They were received by Prof. Fanao Kong from the Chinese Academy of Sciences' Institute of Chemistry. It was a cold winter, and both Ahmed and Dema climbed the Great Wall! They also visited the Forbidden City, Tiananmen Square, and museums. He was very excited about the visit, especially the Chinese history and culture, daily people's lives, and students' education in colleges.

He gave a slide show about the Beijing trip to the group when he came back. He was amazed by the thousands of bicycles orderly moving in one direction, without chaos, every day. I still remember one particular thing he told us: When he visited the famous electronics street in Zhong Guan Cun, he was impressed by many people riding bicycles with big computer boxes on their backseats. He said there is a dream and future in that box.

Nearly after eight years and in summer 2004, I accompanied Ahmed and Dema to visit China again. We spent about two weeks and visited three cities, Beijing, Wuhan, and Shanghai. In Beijing, he received the honorable Doctoral degree from Peking University, a very prestigious honor that only a few scientists have achieved as our host Prof. Qihuang Gong told us. At the awards ceremony, Ahmed had a signing activity for his translated book premiere, *Voyage through Time* (Figure 29.4). This activity continued in all three cities and thousands of students stood in line to get his signature. I was told later that his book has been the most inspirational biography for the young generation in the last 50 years and has

Figure 29.4. Ahmed Zewail signed his translated book, *Voyage Through Time*, in Peking University in June 2004.

influenced millions of young Chinese people. We also had a meeting with President Yongxiang Lu of Chinese National Academy of Sciences. We enjoyed visiting the royal gardens such as the Summer Palace and Beihai Park (Figure 29.5). In a tour to my former Alma Mater, Huazhong University of Science and Technology, in Wuhan, a city that was called one of four ovens (or stoves) in China, we experienced hot weather, upwards of more than 40°C. As many of us know, Ahmed was not a fan of hot weather and it was a huge challenge for me when Ahmed gave a

Figure 29.5. Ahmed and Dema visited Beijing in June 2004.

speech to thousands of students with a big stove fan to blow his back. His shirt was completely wet. I was extremely concerned about him, especially his hour-long book-signing activity after the speech. Luckily, he did not fall ill and I was relieved. The visit to Shanghai was pleasant, and both Ahmed and Dema were very happy to see this "Paris of the Orient" (or Magic City) with both the colonial and modern buildings on the Shanghai Bund. Also, Ahmed gave a public speech at my former school, Fudan University, received by thousands of faculty and students. Ahmed was very impressed by many good questions from the students and their fluent English. During the week, we three visited Yu Garden (Figure 29.6), the Shanghai Museum, New World, Nanjing Road, and many landmarks.

Figure 29.6. Ahmed Zewail with the author in Yu Garden of Shanghai in June 2004.

Both Ahmed and Dema were amazed by the recent development in China and told me they were very fond of Shanghai. When I suggested visiting the neighboring scenic spots like the West Lake and Water Villages, Ahmed promised me to visit Shanghai again in the near future. There was an arrangement in 2009 after the 9th Femtochemistry conference in Beijing, but Ahmed could not make it because he served on PCAST (the President's Council of Advisors on Science and Technology) in the Obama Administration and had a board meeting in the White House. The recent arrangement to visit Shanghai was made on April 29, 2016, but very unfortunately he could not make it due to illness. I was immediately invited to give a memorial open talk about Dr. Ahmed Zewail on the 3rd National Ultrafast Meeting in China on August 20, 2016, and this was probably the first memorial talk about him in a scientific meeting in the world. Chinese people remembered him.

In 2007, I received tenure. Ahmed was so happy and told his family immediately at dinner. He sent me some beautiful gifts and congratulated me for the early promotion. In 2008, we invited him to come to The Ohio State University to give the Robert Ross Lecture in Biophysics and the Alpheus Smith Lecture in Physics to the public on campus, where I bear Alpheus' son, Robert Smith, professorship. Ahmed brought his Nobel pictures with his signature as a gift for my two sons and both boys were truly thrilled, as well as beautiful gifts to me and my wife. He visited our home and we had a wonderful Chinese dinner with Peking duck, one of his favorite Chinese dishes. He gave a great public talk on "Mystery of Time," covering wide knowledge and time from the ancient Egypt, to Galileo, Newton, and Einstein, and to the modern physics. Thousands of people, including many high-school students and my two little boys, attended this popular lecture in town. It was the best lecture in recent years, as many colleagues told me later.

The history between Ahmed and I is interesting. When I joined Caltech in 1993, I wanted to do ultrafast electron diffraction but I moved to femtochemistry. When Ahmed invited me during my postdoc period and the real opportunity for ultrafast electron diffraction came, I was attracted to femtobiology. Later on, Ahmed always joked to me, "You know, Dongping, you missed the big opportunity." In 2010 when I became a full, endowed professor, I took my first sabbatical leave to Caltech and learnt 4D electron microscopy (4D EM) in the Zewail group and again in 2014, I spent my Guggenheim fellowship in his group. I told him that I had come back! He was very generous and fully supported me to build 4D EM in Ohio State if I got the funding, even though the pattern was not released yet. We had many interactions and talked about science, society, politics, the Middle East, and more. He brought me to the famous round table in Athenaeum many times to let me feel the big minds and wisdom of Caltech's most prominent scholars. Of course, we never forgot to have a cigar during some sunny afternoons in his backyard. Over the past 15 years since I left Caltech, Ahmed and I have become close friends and he took me as his brother (or son). We both shared similar values. We talked about ancient Egypt and China, and he enjoyed the Confucian philosophy and wisdom. One holiday season, he sent me a gift, a book on Chinese history. From the depths of my heart, I quote what the ancients in China said: "Once a teacher, always a father."

Even after he was sick in 2013, we kept planning many things together. I continued to visit him several times in 2015, and in late November that year, after Thanksgiving, I visited him and we spent the entire afternoon to discuss all details of the conference in Shanghai and Singapore in 2016 to celebrate his 70th birthday with the theme "Frontiers in Physics, Chemistry and Biology." On February 26, 2016, my wife and I were extremely happy and honored to attend a one-day celebration of Science and Society in honor of Ahmed Zewail's 70th birthday in Caltech and hug him for the last time. We never thought he would leave us so early. I even emailed him around May 2016 to plan a visit to him in July about our changes to the meetings in Shanghai and Singapore from our previously planned May to December. He told me that he would have a vacation with his family at the end of July and would like to see me in late August. I flew to Singapore in early August to fix all arrangements for the meeting in December and he had suddenly left us. But, we will have this meeting for sure, this time, in memory of him. We all met many people in our life — some we remember forever and some we forget instantly. Ahmed belongs to the former. We will remember him; his smile, his laugh, his humor, his charm, his love, his vision… and above all, his spirit. Ahmed, you are with us forever!

Author Biography

Dongping Zhong received his B.S. in Laser Physics from Huazhong University of Science and Technology in China and his Ph.D. in Chemical Physics under Ahmed H. Zewail from California Institute of Technology in 1999. For his Ph.D. work, Dr. Zhong received the Herbert Newby McCoy Award and the Milton and Francis Clauser Doctoral Prize from Caltech. He continued his postdoctoral research in the same group at Caltech with focus on protein dynamics. In 2002, he joined The Ohio State University as an Assistant Professor, and currently he is Robert Smith Professor of Physics and Professor of Chemistry and Biochemistry. He is the Packard Fellow, Sloan Fellow, Camille Dreyfus Teacher-Scholar, Guggenheim Fellow, APS Fellow, AAAS Fellow, as well as the recipient of the NSF CAREER Award and the OCPA Outstanding Young Researcher Award. His research interests include biomolecular interactions and dynamics using ultrafast photon and electron methods.

Chapter 30

Ahmed Zewail — An Inspired and Inspirational Scientist and Man

David Phillips*

In 1986, Sir John Meurig Thomas took up his position as Director of the Royal Institution of Great Britain (Ri), with me as Deputy Director. John liked to keep abreast of advances in all spheres of chemistry, and early in our joint time at the Ri, asked me who in the world best understood the concept of coherence. There was no doubt in my mind that this was Ahmed Zewail, who had published much on observed quantum interference effects in, for example, sodium iodide dissociation and reformation, but who was also exploring with Peter Felker the phenomenon of rotational coherences. I brought Ahmed's work to John's attention, and the rest is history, since a deep friendship developed between the two, which is outlined in the submission in this volume by Sir John.

My interactions with Ahmed were primarily at conferences across the world, and of course a shared interest in photophysics. My own modest contribution to the study of the photophysics of molecules cooled in "supersonic jets" sprang from early work in the gas phase during a

---

*d.phillips@imperial.ac.uk

postdoctoral position in 1964–1966 in the University of Texas working on the fluorescence and nonradiative decay of aromatic molecules with "the grand old man of photochemistry," W. Albert Noyes Jr. At that time, dynamic measurements of fluorescence decay were difficult, but by the late 1960s, time-correlated single-photon counting had developed to the point where nanosecond measurements were possible, the light sources used being low-intensity spark discharge lamps. Nevertheless, for the first time, the dynamics of excited state molecules became amenable to study, prompting, for example, the hugely significant development of the theory of nonradiative decay by Heller, Rice, Jortner, Hochstrasser, and others. However, interpretation of results was complicated by the fact that excitation, even with narrow-bandwidth light sources, populated a huge number of excited state energy levels. It was clear that more useful photophysical information could be gained by "isolating" molecules in the gas phase at low pressures, and many groups published in this area, including our own in Southampton University, where I had become a lecturer in 1967.

The work of Don Levy and others on "supercooling" molecules by supersonic jet expansion opened up a whole field of high-resolution spectroscopy, followed by the use of ultrafast lasers to study the dynamics of decay of excited states that caught the world's attention, with Ahmed Zewail pioneering research on this field with a succession of truly brilliant studies. In 1980, my group moved to the Royal Institution, where among other interests, we were concerned with the study of aromatic molecules "solvated" sequentially by one, two, three, and more solvent molecules to try to unravel the effects of solvent upon the decay dynamics of molecules of interest. We thus became part of a group of like-minded researchers across the world dubbed by Ed Schlag as "The Solvation Army." This area was not Ahmed's prime interest, but he wrote several times to pass critical comment on our papers, and of course we discussed the work at the many conferences we both attended.

Ahmed and Peter Felker made a huge contribution to a volume titled *Jet Spectroscopy and Molecular Dynamics*, edited by John Hollas and myself in 1995. In one chapter, the cause and observation of rotational coherence phenomena on relatively long time-scales was outlined, and the use of this technique to yield structural information demonstrated. Ahmed went on, of course, to seek other means of providing such structural information on excited states in the later brilliant development of electron diffraction on the ultrafast timescale. In the second chapter he and Peter contributed to the volume, ultrafast dynamics of vibrational redistribution in complex polyatomic molecules was covered.

Ahmed was a great admirer of George Porter, developer with Norrish of the technique of flash photolysis, Nobel Laureate in Chemistry in 1967, and Director of the Royal Institution from 1966 to 1985, when he became President of the Royal Society. In a contribution from Ahmed to a volume dedicated to the memory of George Porter edited by myself and Jim Barber published in 2006 titled *The Life and Scientific Legacy of George Porter*, Ahmed reproduced his Nobel Prize Report from *J. Chem. Ed.* in 2001, which contained details of all of the wonderful science he had done. I always admired most the elegant femtosecond study on bond-breaking and reformation in sodium iodide which featured large in this *J. Chem. Ed.* article. In his appreciation of George Porter in our volume, Ahmed wrote:

"The invention of lasers changed the landscape. Flash lamps [used earlier in flash photolysis] did not provide the versatility and intensity of pulsed lasers, and George and his group were among the first to use these pulsed lasers to improve the time resolution. But George was concerned about the limitation imposed by the 'uncertainty principle' on future advances in the picosecond and especially the femtosecond time regime. What was not obvious at the time was the fundamental role of 'coherence'. Even in isolated molecules, coherence can be induced among quantum states, in fact by *exploiting* the uncertainty principle. What is relevant is the energy uncertainty compared to the total energy of the system under consideration. These realisations led to the advances made in studies of ultra-fast molecular phenomena at much shorter timescales; in 1980 with picosecond, and in 1987 with femtosecond time resolution. As importantly, it was possible to distinguish between 'kinetics of ensembles' and 'dynamics of single molecule trajectories'. Without clear understanding of the meaning of coherence, these conceptual limitations would have impeded, or perhaps delayed the impact of ultra-short time- resolution in chemistry and biology."

Another reason for Ahmed's admiration for George Porter was his acknowledgment that Porter was a superb public lecturer, who particularly enjoyed enthusing the young with science, a challenge many of us took on at the home of the "demonstration lecture," the Ri. Ahmed Zewail was himself deeply committed to inspiring young scientists and the lay public, and gave a memorable Friday Evening Discourse at the Ri, as John Meurig Thomas has described in this Volume.

In 1995, Ahmed was one of the ten Plenary Speakers at the International Conference on Photochemistry in Imperial College, of which I was the organizer. Photochemistry was by then a very broad subject, and we divided the scientific proceedings into some ten or so subject areas, of which reaction dynamics and ultrafast reaction was one. Ahmed gave a magisterial account of his recent work in the field. We tried to arrange a varied social program as well as stellar science during that week, which was uncharacteristically for London, very warm, with temperatures upward of 90F for most of the week. Accommodation for delegates was in student dormitories, which needless to say, were not air-conditioned. Ahmed was one of the several delegates who chose to move to nearby hotels for relief, but unfortunately, the one he chose was also not air-conditioned, so the relief was alas very limited!

We arranged the "mixer" for the conference in one of the galleries of the Science Museum, adjacent to Imperial College, and a highlight of the Wednesday afternoon was a boat trip down the Thames with refreshments and live jazz, followed for some delegates by a visit to the penthouse apartment of Mary and Jeffrey Archer for further drinks. Ahmed was one of the 20 or so guests.

The climax of the social events of the week was the Conference Banquet which was held on the ground floor of the Natural History Museum, around the skeleton of the Diplodocus dinosaur (Dippy, as it was affectionately known). The top tables were right under the tail of the skeleton, prompting one speaker to observe this was the first time he had spoken gazing up the fundament of a huge, fortunately extinct, animal. The top table held around 20 guests, and somewhat inexplicably, we had not seated Ahmed at this table (a poor excuse but his Nobel came in 1999); however, he was not deterred by this, and moved himself to be seated next to George

An Inspired and Inspirational Scientist and Man   225

Figure 30.1. Ahmed Zewail giving an impromptu vote of thanks at the Conference Banquet (in the Natural History Museum) of the International Conference on Photochemistry, 1995, held at Imperial College. Ed Schlag and Frank Wilkinson are identifiable in the photograph and, on the far right, George Porter can be seen.

and Stella Porter, and unbidden, seized the microphone during speeches to give an impromptu and much appreciated address (Figure 30.1). He himself made one of the "wonderful contributions" he refers to in his letter to me of August 1995.

I was never a very close friend of Ahmed's, but from the late 1970s onward, met him many times at conferences across the world, and in the UK at the Ri, and Imperial College. Ahmed was never less than hugely courteous to me personally and expressed genuine interest in our research. His own stellar position in our science did not lead to any sense of self-importance as his modest reply to my letter of congratulation on his Nobel Prize in 1999 indicates. He was a legend to my research students and colleagues, and his body of work will be a lasting testimony to one who left us far too soon-a truly great scientist and human being.

CALIFORNIA INSTITUTE OF TECHNOLOGY

Arthur Amos Noyes Laboratory of Chemical Physics, Mail Code 127-72
Pasadena, California 91125

AHMED H. ZEWAIL
Linus Pauling Professor of Chemistry
and Professor of Physics

Telephone: 818-395-6536
Telex: 675425 CALTECH PSD
Fax: 818-792-8456

August 7, 1995

Professor David Phillips, Chairman
Local Organizing Committee
XVIIth International Photochemistry Conference
Department of Chemistry
Imperial College of Science, Technology and Medicine
Exhibition Road
South Kensington
London SW7 2AY, United Kingdom

Dear David:

This is just a note to congratulate you on the ICP Conference. I know what it means to organize such a conference and you should be proud of its success and the wonderful contributions.

With my very best wishes,

Ahmed H. Zewail

AHZ:ms

CALIFORNIA INSTITUTE OF TECHNOLOGY
Arthur Amos Noyes Laboratory of Chemical Physics, Mail Code 127-72
Pasadena, California 91125

Ahmed H. Zewail

March 29, 2000

Prof. David Phillips
Department of Physical Chemistry
Imperial College of Science, Technology and Medicine
London SW7 2AY, United Kingdom

Dear David,

Thank you so much for your thoughtful and kind words of congratulation for the 1999 Nobel Prize. I am very pleased to have the work of our group so recognized, but I must say that the appreciation and support of so many colleagues and friends all over the world constitutes the real award.

With best wishes and happy 2000!

AHZ/sj

Author Biography

David Phillips is a graduate of the University of Birmingham (B.Sc. Chemistry 1961, Ph.D. 1964, Honorary D.Sc., 2011). He carried out post-doctoral work in the University of Texas at Austin 1964–1966 (Fulbright Scholar), and the Institute of Chemical Physics, Academy of Sciences of USSR, Moscow (1966–1967), before beginning his University teaching and research career at the University of Southampton, Department of

Chemistry, in 1967. He then moved to London in 1980 as Wolfson Professor of Natural Philosophy at the Royal Institution and, in 1989, became Hofmann Professor of Chemistry at Imperial College London (1989–2006) and Head of Department (1992–2002). He served as Dean for the Faculties of Life Sciences and Physical Sciences (2002–2005) and Senior Dean (2005–2006) at Imperial. He is the author of some 595 articles and reviews in his field of photochemistry, specializing in photophysics of aromatic molecules and amines, including some jet spectroscopy, and in photodynamic therapy. He received the Porter Medal of the European, Oceanic, and Inter-American Photochemistry Associations 2010 for this work. He is a committed popularizer of science and was awarded the Michael Faraday Award of the Royal Society in 1997 and an OBE in 1999 for services to science education, and CBE in the 2011–2012 New Year's honors list for services to Chemistry. He was President of the Royal Society of Chemistry (2010–2012) and elected Fellow of the Royal Society in 2015.

Chapter 31

My Memories of Ahmed Zewail — From a Snowy Northern Sweden to the Nobel Prize

Villy Sundström*

In the summer of 1993, two Russian scientists from the Institute of Physics in Nizhny Novgorod drove the 1,934 km to Umeå, where I was based at that time, in their red Lada car loaded with a home-built Ti:Sa laser. Within a few days, the laser was put in operation, pumped by a small argon-ion laser, and produced stable ~100 fs pulses at 790 nm. These pulses were applied to our favorite topic at that time, energy transfer within light-harvesting complexes of photosynthetic purple bacteria. Energy transfer times between the pigment molecules of two different LH2 complexes were measured and the work was published in *Chem. Phys. Lett.* (*CPL*).[1] In March 1994, Ahmed was touring Sweden, and having seen our recently published LH2 work in *CPL*, he took the detour to northern Umeå. On March 24, I picked up Ahmed outside the "Stora Hotellet" in Umeå, dressed in a thin summer jacket and shoes suitable for California climate, but not for a cold, snowy, and gloomy day in March in

---

* villy.sundstrom@chemphys.lu.se

northern Sweden. This was how I met Ahmed for the first time. We quickly moved to the chemistry department, where we had a full day of scientific discussions. We had continued the work on energy transfer in photosynthetic light-harvesting complexes and had a bunch of exciting new results that we thought indicated vibrational coherence and inter-exciton state relaxation processes. Ahmed showed his enthusiasm and support to the results, which we submitted to *CPM* a month later.[2,3] Through this work, and much work that followed as well as work by other groups, a deep understanding of the primary processes of photosynthesis was reached (see, e.g, Ref. 4). Nevertheless, using new sophisticated ultrafast methods such as 2D electronic spectroscopy, increasingly more mechanistic detail is revealed still today (see, e.g., Ref. 5). At the dinner we had later in the evening, Ahmed encountered a northern specialty, which I think became somewhat of a favorite dessert — warm cloudberries with vanilla ice cream.

When Ahmed visited Umeå, it was already clear that I would move to Lund later that year to start up a new operation. I told Ahmed about the plans to organize a small farewell workshop on some venue characteristic for northern Sweden. Ahmed gave the idea full support and the location that was eventually chosen was a small island, Ulvön (Figure 31.1), in the Baltic Sea, approximately 150 km south of Umeå and 500 km north of Stockholm. Many of the key players of the ultrafast community at that time were invited, and during the first days in June 1994, we had a fantastic conference when participants also could enjoy the bright northern summer nights and the fantastic nature. The island of Ulvön is famous for being the home of a very special delicacy — "surströmming," which is fermented herring. You eat it with the typical northern Sweden thin crispy bread, potatoes, onion, and perhaps a little bit of sour cream, and it goes best with beer and a "snaps." The conference dinner offered a sample of this delight and most conference participants were brave enough to try — Ahmed belonged to the strong supporters of surströmming.

After moving to Lund in summer of 1994, I started to build the Chemical Physics Division. At the start, we were five lonely souls in the big Chemistry Department, but we quickly grew when students and postdocs from all over the world joined. The first Ph.D. student to graduate in

Figure 31.1. View over Ulvön, where the June 1994 workshop took place.

1997 was Mirianas Chachisvilis, who had been working on photosynthetic energy transfer. Ahmed was his Thesis faculty opponent, and after a record-quick questioning and examination of the Thesis, Mirianas ended up as a postdoc in Pasadena. Jennifer Herek came from the Femtoland at Caltech, and the world of unimolecular reactions of small molecules, to Lund and energy transfer dynamics of photosynthetic proteins. A highlight of Jennifer's work in Lund was the successful coherent control of energy transfer in a LH2 antenna complex of photosynthetic purple bacteria,[6] a work she performed in collaboration with Marcus Motzkus, another young scientist from the Zewail family.

In 1996, I had the great privilege to organize a Nobel Symposium on Femtochemistry and Femtobiology. Twenty-five leading scientists were invited to the symposium as lecturers and discussion leaders, and about an equal number of young researchers were invited as observers. The scientific program consisted of 19 lectures, which are all published in a proceedings volume.[7] The symposium took place at Hotel Svartå Herrgård and Alfred Nobel's Björkborn Manor. In order to set the right atmosphere, the symposium opened with a reception at the Royal Swedish Academy of Sciences in Stockholm, where the participants were informed on the work of the Academy and had the opportunity to see the session room where the final voting of the Nobel Prizes take place. The conference dinner was a particularly memorable event, when, while waiting for the dessert, "Alfred Nobel himself" entered the room and started chatting with the participants about his own research in the field of explosives and inquired about work and presentations of the participants. Still many years after his encounter with "Alfred Nobel himself," Ahmed talked about it as one of the highlights of the symposium. In the early days of the Nobel symposia, they had a screening function for a possible Nobel Prize in a particular field. In 1996, with the advent of the internet, much more frequent publishing and easier access to published results, the symposia no more had this function. Nevertheless, it was clear that all participants invited accepted the invitation, took it very seriously, and very good science was presented and discussed at the many intermissions.

The Nobel Prize in 1999 was, also for me, an unforgettable highlight. In connection with each Prize, the Swedish Royal Academy of Sciences (Kungliga Vetenskapsakademien, KVA) each year produces a poster to illustrate the essence of the Prize in a popular way. I had the pleasure to be part of this process for Ahmed's Prize (Figure 31.2), together with Eva Krutmeijer of the Academy and specialists in graphic design. The focus of attention was, of course, an illustration of one of the reactions studied by Ahmed and his group, with reactants, the Transition State Structure, and products connected via their potential energy surface. It was natural for us to get connections to "the Swedish hero" Svante Arrhenius. I remember that in an early stage of the poster, Arrhenius sitting on his cloud, was holding the Transition State double dagger in his hand. In the final version of the poster that detail disappeared, looking too much like the religious symbol of Christianity. We also made connections to the origin of high-speed photography through Muybridge's galloping horses, as well as the concept of "Molecular Movies," which only now is becoming true reality

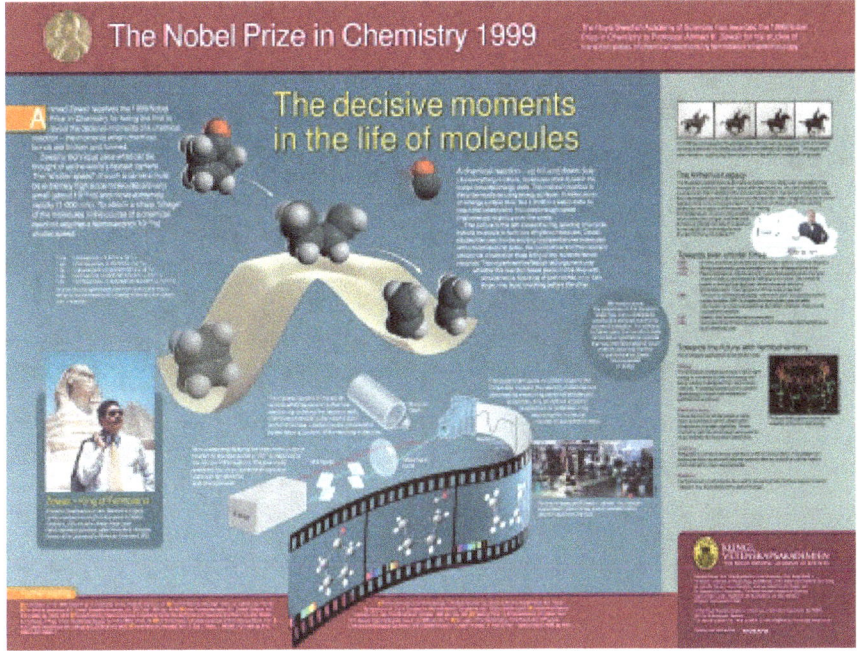

Figure 31.2. The Nobel Poster illustrating the Nobel Prize in Chemistry 1999. (© The Royal Swedish Academy of Sciences, Graphic design: Kjell Lundin, Explicare.)

through ultrafast electron diffraction and microscopy and ultrafast X-ray methods. Finally, we could not resist placing the "King of Femtoland" in the habitat of his childhood with the Sphinx in the background.

Following the Nobel festivities in Stockholm, Ahmed visited Lund and gave a much-appreciated lecture. The big lecture Hall A at the Chemistry Department was filled as never before, and he was received like a rockstar by the Egyptian community in Lund. Once more, Ahmed had the pleasure to enjoy his favorite dessert, warm cloudberries and vanilla ice cream, at the restaurant in Grand Hotel.

Another memorable event was when Ahmed was promoted to honorary doctor (Doctor *Honoris Causa*) at Lund University in 2003 (Figure 31.3). On a beautiful sunny day in May, newly baked doctors, honorary doctors, guests, and university officials dressed up in gowns,

Figure 31.3. In May 2003, Ahmed was promoted to Doctor *Honoris Causa* of Lund University. (Photo and copyright: Kennet Ruona).

black tails, and long dresses gathered in the university administration building for the short procession to the cathedral where the procedure took place. After more than three hours sitting on medieval wooden benches, listening to the promotion speeches in Latin, and with aching backs, we were all happy to get out into the sun again and the waiting reception.

I would like to thank you Ahmed for your inspiration and support through the years, from general attitude to the scientific mission to our joint work as *Chem. Phys. Lett.* editors.

References

1. S. Hess, F. Feldshtein, A. Babin, E. Åkesson, T. Pullerits and V. Sundström, "Femtosecond Energy Transfer within the LH2 Peripheral Antenna of Photosynthetic Purple Bacteria, *Rhodobacter sphaeroides* and *Rhodopseudomonas palustris*," *Chem. Phys. Lett.* **216**, 247–257 (1993).
2. M. Chachisvilis, T. Pullerits, M.R. Jones, C.N. Hunter and V. Sundström, "Vibrational Dynamics in the Light Harvesting Complexes of the

Photosynthetic Purple Bacterium *Rhodobacter sphaeroides*," *Chem. Phys. Lett.* **224**, 345–354 (1994).
3. T. Pullerits, M. Chachisvilis, M.R. Jones, C.N. Hunter and V. Sundström, "Exciton Dynamics in the Light Harvesting Complexes of *Rhodobacter sphaeroides*," *Chem. Phys. Lett.* **224**, 355–365 (1994).
4. V. Sundström, T. Pullerits and R. van Grondelle, "Photosynthetic Light-harvesting: Reconciling Dynamics and Structure of Purple Bacterial LH2 Reveals Function of Photosynthetic Unit," *J. Phys. Chem.* **103**, 2327–2346 (1999).
5. J. Dostál, J. Pšenčík and D. Zigmantas, "In Situ Mapping of the Energy Flow through the Entire Photosynthetic Apparatus," *Nat. Chem.* **8**(7), 705–710 (2016).
6. J.L. Herek, W. Wohlleben, R.J. Cogdell, D. Zeidler and M. Motzkus, "Quantum Control of Energy Flow in Light Harvesting," *Nature* **417**, 533–535 (2002).
7. V. Sundström, *Femtochemistry and Femtobiology — Ultrafast Reaction Dynamics at Atomic-Scale Resolution*, Nobel Symposium 101, Björkborn, Sweden, September 9–12, 1996 (Imperial College Press, London, 1997).

Author Biography

Villy Sundström received his Ph.D. at Umeå University, Sweden, in 1977, after studies at Bell Laboratories under the guidance of Prof. Peter Rentzepis. At Umeå University, he later built the first picosecond laboratory in Scandinavia. In 1994, he moved to Lund University, where the Chemical Physics Division was created, which today houses approximately 50 scientists and students working in six laser laboratories for ultrafast and single molecule spectroscopy. Sundström received an ERC Advanced Investigator Award 2008, is an Editor of *Chemical Physics Letters*, and Member of the Royal Swedish Academy of Science. His research interests include:

- Excited state and charge carrier dynamics in nanostructured materials for solar energy conversion.
- Chemical reaction dynamics.
- Ultrafast structural dynamics in chemical and biological systems, studied with time resolved X-ray spectroscopy.

- Photophysics and photochemistry of melanin and other natural pigments and their building blocks.
- Femtobiology: Ultrafast spectroscopy applied to various biological systems.
- Photosynthetic light-harvesting. Energy flow pathways and energy transfer mechanisms.

Chapter 32

The Physical Basis of the Amyloid Phenomenon

*Christopher M. Dobson**

Abstract

Interest in the phenomenon of amyloid formation by peptides and proteins has developed with extraordinary rapidity in recent years, to the extent that it is now a major topic of research across a wide range of disciplines. This surge of interest was initially prompted as a result of the links between amyloid formation and a range of rapidly proliferating medical disorders that include Alzheimer's disease and Type-2 diabetes. More recently, it has been recognized that an understanding of the amyloid phenomenon provides fundamental insights into the inherent nature of the biologically functional forms of peptides and proteins and the means of the maintenance of protein homeostasis within healthy living systems. In addition, the characteristics of the generic amyloid structure have led to its investigation as the basis for a wide range of biomaterials with fascinating and widely tunable properties. Recent progress in understanding the factors affecting the stability of the amyloid state relative to that of the native state of a protein, along with the development of methods for defining the mechanism of the conversion between the

---

*cmd44@cam.ac.uk

different states, has led to a much more detailed understanding of the nature of the amyloid phenomenon, and this chapter, dedicated to Ahmed Zewail, provides a summary of some recent advances in this field of study.

In 1906, Alois Alzheimer gave a lecture in which he discussed the case of one of his patients in the Frankfurt Asylum, named Auguste Deter. Deter was just over 50 years old and was experiencing unfamiliar symptoms, including cognitive failure and memory loss. This was the first description of the disorder now known as Alzheimer's disease (AD), and after Deter's death, Alzheimer examined her brain and found that it contained a mass of deposits or plaques known as "amyloid" because they stained with dyes that also stain for starch, which in Latin is "amylum." Such "starch-like" deposits are, however, not composed of carbohydrates but of proteins, which in the case of Alzheimer's disease is primarily a 42-residue fragment (known as the amyloid-β peptide, or simply Aβ) of a large transmembrane protein, the amyloid precursor protein (APP). Moreover, when viewed under the electron microscope, these deposits can be seen to contain thread like "amyloid fibrils," which are typically some 10 nm in diameter but can range up to microns in length.

Since that first description of a patient suffering from an unfamiliar condition, the number of people today with AD is estimated to be about 40 million, spread across the world.[1] The dominant risk factor leading to AD is age (although we now know by analysis of preserved tissue that Auguste Deter was suffering from a rare early-onset familial form of AD) and this huge increase in the number of cases is attributable to the dramatic increase in longevity that has occurred over the last century that has resulted from greatly enhanced sanitation and public health measures along with tremendous increases in medical science that have, in particular, dramatically reduced the number of cases of infectious diseases, notably in the most affluent parts of the world, ranging from bubonic plague to smallpox. The major causes of death in much of the world are now non-infectious, often chronic, diseases such as heart failure, cancer, and increasingly, dementia, of which 80% of cases are AD. Like AD, the majority of these chronic diseases are age-related, but AD is particularly dramatic in this context as the probability of suffering from this condition is 1%–2% at the age of

65 years, but rises to 30%–50% at the age of 85 years. The number of people in the world aged above 65 years is expected to triple in the next 35 years, and by 2050, it is estimated that over 130 million people will have AD, over 70% of whom will be living in low- or middle-income countries.[1] Because of the nature of AD, with many patients living for 10 years or more but requiring continuous care, the cost and social consequences of this disorder are enormous. The other reason for the dramatic rise in the numbers of AD patients is that the condition is incurable and indeed untreatable, except for the amelioration of some of its symptoms.

One of the other remarkable features associated with the appearance of thread-like amyloid fibrils in the brains of AD patients is that more than 50 other disorders are now known to be linked with the appearance of such structures in organs and tissues, all of which are currently incurable.[2] These disorders include other neurodegenerative conditions such as Parkinson's and Huntington's diseases, and others where amyloid deposition occurs in organs other than the brain, such as the heart, liver, and pancreas. The latter is of particular interest as such deposition is linked to Type-2 diabetes, a condition that is estimated to affect over 400 million people around the world. The amyloid deposits associated with each of these disorders involve a specific dominant protein, but there are no simple correlations of the various proteins involved in this family of disorders with their sequences, native-state structures or functions, or indeed why or how these proteins converted from their functional native folds into these aberrant misfolded structures. Some 20 years ago, however, we observed that proteins with no links to any amyloid disease could convert under laboratory conditions into fibrillar forms with all the features of the amyloid structures extracted from patients suffering from one or other of these "misfolding diseases".[3] It soon became apparent that the amyloid state is a generic alternative form of protein structures whose similar overall architecture is determined by the interactions within the main chain that is the same for all proteins, whilst the myriad array of distinctive native folds is fundamentally defined by the interactions between the side-chains that are defined by the unique amino acid sequence of each type of protein (Figure 32.1).[4]

In recent years, tremendous progress has been made in understanding the nature, properties, and significance of the amyloid state of proteins.

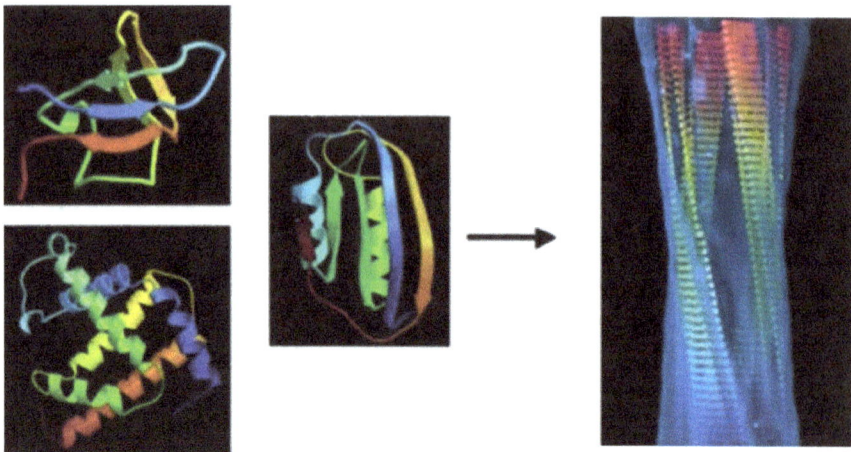

Figure 32.1. Comparison of examples of native and amyloid structures of protein molecules. On the left are ribbon diagrams of the native structures of three small proteins: an SH3 domain (top), myoglobin (bottom), and acylphosphatase (middle). The native structures differ in their topologies and contents of α-helices and β-sheets resulting from the dominance of side-chain interactions within their highly evolved sequences. On the right is a molecular model of an amyloid fibril (image kindly provided by Helen Saibil, Birkbeck College, London). The fibril was produced from the SH3 domain whose native structure is shown on the left, and consists of four "protofilaments" that twist around one another to form a hollow tube with a diameter of approximately 6 nm. The β-strands (flat arrows) are oriented perpendicular to the fibril axis and are linked together by hydrogen bonds involving main chain amide and carbonyl groups, many of which are intermolecular, to form a continuous structure in each protofilament. The protofilaments are held together by much weaker interactions involving primarily side-chain contacts. As the main chain is common to all polypeptides, the core protofilament structures of fibrils from different sequences have common features, differing only in detail as a result of differences in the nondominant effects of side-chain packing. The arrow indicates that when the native states of globular proteins are destabilized, they tend to convert into the generic amyloid structure, as described in the text.

Source: From Ref. 4.

Although it has long been known from X-ray fiber diffraction studies that amyloid fibrils have a "cross-β" structure in which β-strands are oriented perpendicularly to the fibril axis, the structures of several fibrils are now known in atomic detail, showing the common occurrence of a closely interacting pair of β-sheets with a slight twist (Figure 32.2).[5] Such

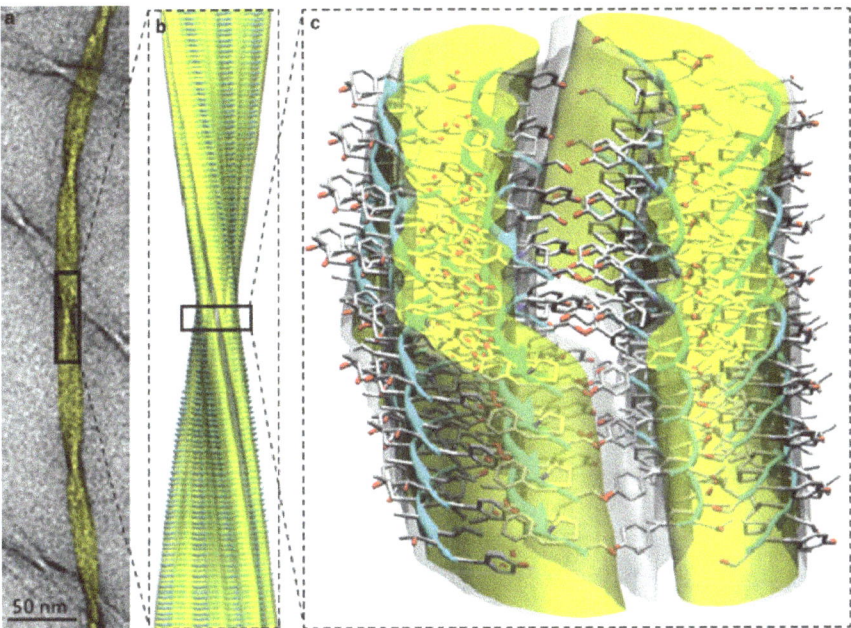

Figure 32.2. Structure of an amyloid fibril at atomic resolution. The structure shown is one of several polymorphs of the amyloid fibrils that are formed from an 11-residue fragment of transthyretin. The combination of cryo-electron microscopy imaging (part a) with solid-state NMR analysis has enabled the determination of an atomic-level structure (part b). A more detailed view (part c) shows the hierarchical organization of the amyloid fibril in which the three filaments that form the mature fibril illustrated here are in turn formed by pairs of cross-β protofilamants, which are each composed of pairs of β-sheets. The fibril surfaces are shown as electron density maps, and the constituent β-sheets are shown in a ribbon representation; oxygen, carbon, and nitrogen atoms are shown in red, gray and blue, respectively.

Source: From Ref. 5.

structures confirm the importance of arrays of hydrogen bonds linking amide and carbonyl groups of the polypeptide main chain, and although the essential architecture of the amyloid structure is defined by the properties of the main chain, the side chains affect the details of the alignment and spacing of the β-sheets and the regions of the polypeptide chains that are most likely to form the core structures of the fibrils. Of particular importance, however, is the fact that the propensity of a protein to convert from its functional soluble state into the amyloid state depends on its amino acid sequence. This issue turns out to be of considerable significance because of another remarkable and unexpected characteristic of the amyloid state, namely that in many cases (particularly in the case of small proteins or protein fragments), this form of a protein is more stable than the native functional state, even under physiological conditions.[2]

This finding is of very considerable significance as it reveals that the native states of many proteins are metastable, and they are therefore able to remain in their functional forms in living systems only because the kinetics of the process of aggregation and amyloid formation is slow on our biological timescales. In the context of the onset of disease, therefore, the development of a detailed understanding of the factors that determine the kinetics and mechanisms of the processes involved in converting from their soluble states into intractable amyloid fibrils assumes major importance in understanding the onset and progression of disease. In recent years, considerable progress has been made in this area of activity through the development of methods of kinetic analysis that enable the various steps in the aggregation process to be analyzed in detail.[2] Such studies have shown the importance of secondary processes — such as the catalytic nucleation that takes place on fibril surfaces – and also external factors — such as the role of membrane surfaces in stimulating aggregation. They have also revealed the importance of protective mechanisms, such as the many forms of molecular chaperones, within living systems that can inhibit the various steps involved in aggregation, and also provided key information on the origins of the collapse of protein homeostasis that can occur when one such protective mechanism becomes saturated and overwhelmed.

Another key observation in the context of disease is that the cellular toxicity associated with aggregation and amyloid formation does not appear to be primarily due to the mature fibrils and plaques that were initially thought to be the most damaging processes in AD and other conditions, but instead due to the prefibrillar species that are often known as oligomers. It is interesting in the context of the generic nature of the structures and means of formation of amyloid fibrils that such oligomeric species can be highly toxic even when formed from proteins with no connection with disease. The oligomeric species have a high surface-to-volume ratio that promotes aberrant interactions involving the misfolded aggregate surfaces, for example, with cellular membranes, leading to cellular damage and even death. In addition, such species are able to diffuse rapidly from cell to cell, resulting in the transmission of damaging species through mechanisms that are often called "prion-like" by analogy to the mechanisms thought to be responsible for transmissible misfolding disorders such as Creutzfeldt–Jakob disease. Structural studies to define the nature of these oligomers and their aberrant interaction with cellular components such as membranes are now beginning it to bear fruit, despite the challenges of heterogeneity of such species and their frequently transient nature. Indeed, the ability to link together the kinetic events associated with key steps in the process of protein aggregation with the species that are formed at the different stages of this process is leading to an increasingly coherent description of both normal and aberrant biological behavior.[2]

Indeed, one of the very exciting aspects of work in this field of science is the manner in which it reveals which complex biological phenomena can be related to the physical principles that underlie their behavior. It was this aspect of the topic that appealed to Ahmed Zewail and led to many fascinating discussions. I had met Ahmed from time to time over many years, but it was when I moved to Cambridge in 2001 that I got to talk to him regularly, often in the company of John Meurig Thomas, Amyand David Buckingham, and Bengt Nordén. One of my particular memories of Ahmed, however, was at a Welch Foundation meeting in Houston that he had organized in 2007. One of the activities that was just beginning at that time in my research group, and is now of major significance, was concerned with the relationship between the behavior that were able to study in detail in the "test tube" and that occurring in living systems. I remember showing a slide of some experiments that we had been doing with transgenic fruit flies in which the Aβ-peptide associated with AD was expressed.[4] These flies showed a number of characteristics of AD, including the appearance of amyloid deposition, the loss of normal mobility, and reduced longevity. We had then incorporated single mutations within the sequence that from our laboratory studies we predicted would very slightly increase or decrease the propensity of the Aβ-peptide to aggregate.

The results were fascinating, as these quite subtle mutations had very considerable effects on the mobility and lifespan of the flies. Moreover, there was an astonishingly high correlation between the change in aggregation propensity and the functional viability of the flies (see Figure 32.3).[4] Ahmed immediately asked for a copy of the movie we had

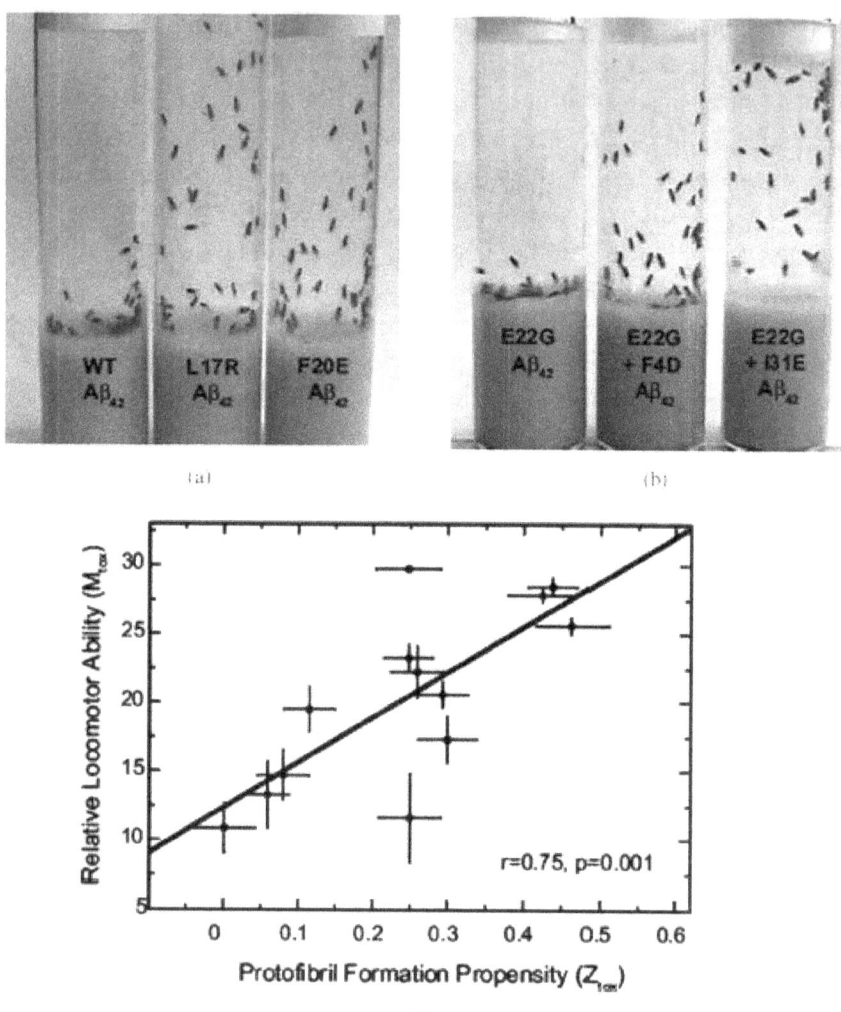

Figure 32.3. The effect of mutations in the sequence of the 42-residue human Alzheimer Aβ-peptide on neuronal dysfunction in transgenic fruit flies. The upper left panel (a) illustrates a climbing assay of flies expressing the wild-type sequence (left) and two mutational variants predicted to reduce the peptide's aggregation propensity; the more mobile the flies the higher up the tube they can climb. The right hand upper panel (b) represents a similar experiment with flies expressing the Aβ-peptide containing the E22G "Arctic mutation" (left-hand tube). The two left-hand tubes are of peptides that contain mutations that decrease the propensity to form prefibrillar aggregates (protofibrils) in particular. The lower panel (c) shows the degree of correlation between the relative locomotor activity of a series of mutational variants against their predicted propensities to form protofibrils.
Source: From Ref. 4.

made to illustrate these findings as he was very excited by the demonstration of an almost perfect correlation between fundamental physics and the biological response of a living organism! Indeed, his enthusiasm to develop an understanding of the ways in which the principles of the physical sciences contribute to our ability to interpret complex phenomena in biology has been a very important factor in encouraging us to pursue such ideas further. Indeed, it was not long after this meeting that we found that it is possible to analyze the kinetics of protein aggregation by means of principles that are similar to those used to analyze the kinetics of simple chemical processes, the general field in which Ahmed had made the seminal contributions that led to his Nobel Prize.

Another consequence of Ahmed's increasing interest in the phenomenon of amyloid formation, and particularly its time dependence, was the idea of using his techniques of time-dependent cryo-electron microscopy (cryo-EM) to explore the relationship between the structure and dynamics of amyloid fibrils. We had already carried out a number of structural studies in this area, including the definition of a high-resolution structure of an amyloid fibril (Figure 32.2).[5] This work had involved a combination of biophysical methods including the use of solid-state NMR spectroscopy and conventional cryo-EM techniques. This work had been the subject of the Ph.D. studies in Cambridge of an extremely talented student, Anthony Fitzpatrick, who then decided to join the Zewail laboratory at Caltech to follow up the idea of applying the technique of time-dependent cryo-EM to the fibrils whose structure had just been determined. This work led to a variety of extremely interesting studies in the Zewail laboratory of the application of 4D cryo-EM to probe the nature of the amyloid structure, demonstrating the potential of such methods in this field of science (Figure 32.4).[6]

Our increasing knowledge of the underlying factors that determine the structure and stability of the amyloid forms of proteins and the processes through which they are formed from normally soluble states of proteins, provides opportunities to intervene in these processes and, in particular, to inhibit the steps that can lead to pathogenicity. In particular, we have focused on the identification of both small and large molecules that can influence specific microscopic steps in the overall mechanisms of amyloid formation.[7] In addition, the accessibility of the amyloid state in conditions other than disease has emerged from the discovery of functional forms of

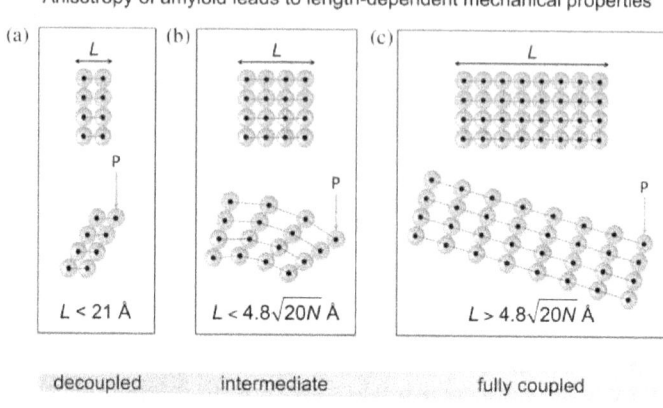

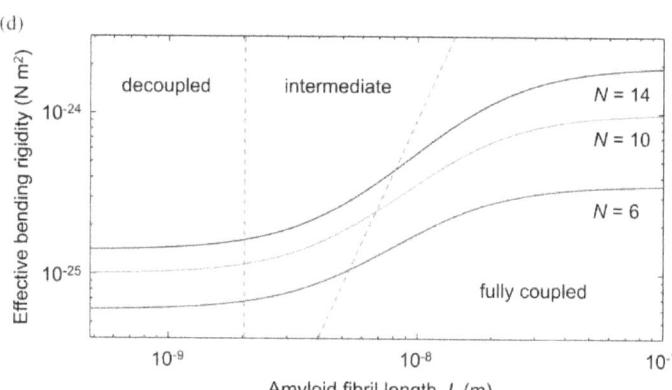

Figure 32.4. Mechanical anisotropy of amyloid leads to length-dependent material properties. An amyloid fibril is a network of rigid β-strands (colored spheres) interconnected via elastic (strong) longitudinal (magenta-dashed lines) and (weak) lateral bonds (yellow-dashed lines). Amyloid fibrils of different lengths, L, along the hydrogen-bonding axis are shown schematically as two laterally connected protofilaments (green and blue spheres represent the first and second protofilament, respectively). The results show that fibrils of different lengths bend through different mechanisms involving bond shearing and extension or compression of intersheet and interprotofilament bonds, and show a correlation between the effective bending rigidity as a function of fibril length.

Source: From Ref. 6, in which further details of these experiments can be found.

amyloid structures in a variety of organisms including humans, being involved in process ranging from catalysis to storage.[8] Another remarkable discovery was the finding that structures with properties related to those of amyloid fibrils could be generated from much smaller units, such as dipeptides, many of which have remarkable chemical and physical properties such as optical and electronic effects that could be extremely valuable in the development of new types of device.[9] Moreover, the ability to convert polypeptide chains in the laboratory into amyloid structures has led to the exploration of their value in materials science, with the vast number of possible sequences leading to an effectively limitless variation in properties.[10]

Ahmed Zewail made seminal contributions to our understanding of molecular dynamics and reaction mechanisms through the development and application of innovative physicochemical methods. His encouragement of the translation of such approaches to biology was invaluable, and the field of science associated with the self-assembly of biological molecules has not only benefited enormously from this philosophy but also led to the recognition that new and surprising physicochemical phenomena can emerge from the study of biological systems. The investigation of the physical principles underlying the onset of some of the most debilitating diseases of the modern world, therefore, promises not only to lead to new and rational approaches to their treatment but also to enable the development of novel types of biologically inspired materials and devices with a range of new and valuable applications in the world of the future.

References

1. Alzheimer's Association, "Alzheimer's Disease: Facts and Figures," (2015).
2. T.P.J. Knowles, M. Vendruscolo and C.M. Dobson, "The Amyloid State and Its Association with Protein Misfolding Diseases," *Nature Rev. Mol. Cell Biol.* **15**, 384–396 (2014).
3. C.M. Dobson, "Protein Misfolding, Evolution and Disease," *Trends Biochem. Sci.* **24**, 329–332 (1999).
4. C.M. Dobson, "Protein Folding and Misfolding: From Atoms to Organisms," in *Physical Biology: From Atoms to Molecules*, A. Zewail (Ed.), (Imperial College Press, London, 2008), pp. 267–279.

5. A.W. Fitzpatrick, G.T. Debelouchina, M.J. Bayro, D.K. Clare, M.A. Caparoni, V.S. Bajaj, C.P. Jaroniec, L. Wang, V. Ladizhansky, S.A. Muller, C.E. MacPhee, C.A. Waudby, H. Mott, A. de Simone, T.P.J. Knowles, H.R. Saibil, M. Vendruscolo, E. Orlova, R.G. Griffin and C.M. Dobson, "Atomic-resolution Structure of a Cross-β Amyloid Fibril," *Proc. Natl. Acad. Sci. USA* **110**, 5468–5473 (2013).
6. A.W.P. Fitzpatrick, G.M. Vanacore and A.H. Zewail, "Nanomechanics and Intermolecular Forces of Amyloid Revealed by Four-dimensional Electron Microscopy," *Proc. Natl. Acad. Sci. USA* **112**, 3380 (2015).
7. J. Habchi, S. Chia, R. Limbocker, B. Mannini, M. Ahn, M. Perni, O. Hansson, P. Arosio, J.R. Kumita, P.K. Challa, S.I. Cohen, S. Linse, C.M. Dobson, T.P.J. Knowles and M. Vendruscolo, "Systematic Development of Small Molecules to Inhibit Specific Microscopic Steps of Aβ42 Aggregation in Alzheimer's Disease," *Proc. Natl. Acad. Sci. USA* **114**, E200–E208 (2016).
8. D.M. Fowler, A.V. Koulov, C. Alory-Jost, M.S. Marks, W.E. Balch, and J.W. Kelly, "Functional Amyloid Formation Within Mammalian Tissue," *PLoS Biol.* **4**, e6 (2006).
9. M. Reches, and E. Gazit, "Casting Metal Nanowires Within Discrete Self-Assembled Peptide Nanotubes," *Science* **300**, 625–627 (2003).
10. T.P.J. Knowles and R. Mezzenga, "Amyloid Fibrils as Building Blocks for Natural and Artificial Functional Materials," *Adv. Mat.* **28**, 6546–6561 (2016).

Author Biography

Chris Dobson is the John Humphrey Plummer Professor of Chemical and Structural Biology at the University of Cambridge, and Master of St John's College. He was an undergraduate and graduate student and research fellow at the University of Oxford. He then became an Assistant Professor of Chemistry at Harvard University before returning to Oxford, where he was a Professor of Chemistry until moving to Cambridge in 2001. His research activities are primarily concerned with understanding the manner in which protein molecules usually fold to generate biological function but can misfold and give rise to disease. His current interests are focused in particular on defining the fundamental origins of neurodegenerative disorders such as Alzheimer's and Parkinson's diseases, with the objective of identifying new strategies for their prevention or treatment. He has published over 750 papers and review articles and is Director of the Cambridge Centre for Misfolding Diseases and a Founder of Wren Pharmaceuticals. He is a Fellow of the Royal Society, the Royal Society of Chemistry, and the Academy of Medical Sciences, and a Foreign Associate of the US National Academy of Sciences. He is the recipient of numerous awards, including the Davy Medal and the Royal Medal of the Royal Society, the Heineken Prize for Biochemistry and Biophysics from the Royal Netherlands Academy of Arts and Sciences, and the Feltrinelli International Prize for Medicine from the Accademia Nazionale dei Lincei in Rome.

Chapter 33

High-Intensity Mentoring and Excitement from Ahmed Zewail

*Marcos Dantus**

It is often difficult to appreciate what one has in the present, and only in retrospect, when everything is given proper perspective, can one realize the true value of a certain event or opportunity. And yet, when I think about my time at Caltech with Ahmed, I honestly knew since I first met him that he was the most inspiring scientist I had ever met. Throughout my eight years at Caltech, during multiple occasions, I knew we were working on the most significant research projects in physical chemistry and that their implications would be worthy of a Nobel Prize.

The series of three papers published by Zewail and coworkers on jet-cooled anthracene had a tremendous impact on me. I was working as an undergraduate along a similar scientific vein; that of elucidating intramolecular vibrational energy redistribution (IVR) in isolated organic molecules. The combination of frequency and picosecond time-resolved spectroscopy served to illustrate the process of IVR, in particular, in the

---

*dantus@chemistry.msu.edu

less-understood restricted regime. When I read those articles and observed the quantum beats among different vibrational states, I had to interview at Caltech and try to be admitted as a graduate student in Zewail's research group.

Once accepted to Caltech, I interviewed with Zewail in 1985. The most exciting moment of that day was when Zewail drew on a napkin the phenol molecule and explained that everyone knows ultraviolet light cleaves the bond and releases the hydroxyl radical. He then looked at me intensely and remarked "but, no one knows how long it takes to break that chemical bond!" He told me it was his dream to measure that process. I looked at him with enthusiasm and said, "I would love to have the opportunity to build the laboratory where such measurements would be carried out."

My first projects at Caltech continued work on IVR in isolated organic molecules, and the role of solvent molecules on IVR and charge transfer processes. In the summer of 1986, Zewail asked me to design a femtosecond chemistry laboratory. The construction of the laboratory progressed at an ultrafast pace, the only pace Zewail approved of. The empty room was ready on November 25, 1985; by the following January, Dr. Mark Rosker (postdoc hired that Fall to help in the construction of the femtosecond laser) and I obtained femtosecond pulses, and we started preparing the necessary wavelengths to carry out the first experiment.

The first experiment was carried out on cyanogen iodide (I–CN), a simple tri-atomic molecule that, upon irradiation with ultraviolet light, released the cyanide radical. Our expectation was to observe the delayed raise in the signal as we probed for the emergence of the cyanide radical. The much-anticipated observation occurred in February, when Mark and I returned from lunch. We started the scan and followed every dot as it appeared on the black computer screen. As the first scan was completed, it became clear it was not as simple as we had anticipated. The signal contained important information regarding the bond-breaking process that had never been observed before. Mark and I started jumping and rushed to call Ahmed to come to the lab.

The writing of the first scientific reports on our findings was not easy. We worked long hours and had to repeat the measurements multiple times with multiple different wavelengths. I learned an important lesson in science: every aspect of an experiment, from calibration of the measurement instruments to the purity of the sample, must be checked thoroughly. The last sentence of the first communication was prescient: "In summary, these FTS experiments promise to provide real-time dynamics of TS and intermediates in unimolecular and special bimolecular reactions."

With a state-of-the-art laboratory for femtosecond chemistry, there was no shortage of projects. Zewail divided us in pairs and we set out to explore ever-increasing complexities in chemical reactions. Work in Zewail's group during this time was so exciting that it was hard to leave the lab and go home. Soon Zewail attracted amazingly talented graduate students and postdocs, and I had wonderful collaborations with them.

As I completed my Ph.D. thesis and was looking for a research group to carry out my postdoctoral work, the idea of performing femtosecond time-resolved electron diffraction came up. I jumped at the opportunity to build a second laboratory. This project proceeded at the same ultrafast speed and achieved time-scales and sensitivities that were considered several orders of magnitude beyond what would be possible. Following that work, it was time for me to start a research group on my own.

My interactions with Zewail as a fellow professor were few but very much appreciated. I could always count on meeting with Zewail at the Femtochemistry conferences every other year. I recall with great pride our meeting in 2003, when the conference was held in Paris that Zewail told me he was particularly impressed with the progress I had achieved using shaped laser pulses.

Zewail always wanted me to organize a Femtochemistry conference in Mexico. He was keenly aware that scientific opportunities in developing countries are limited and knew such a conference would be special. He also wanted to recognize the work of his two Mexican group members, Jorge Peon and myself. Jorge and I organized the Femtochemistry 13 conference in Cancun, Mexico, and Zewail was excited that this would finally take place. Unfortunately, he will be joining us only in spirit.

When I heard of Zewail's passing I was shocked. My consolation was that I had recognized him as an inspiring scientist early on and had the opportunity to work in his group both as a graduate student and then as a postdoc. I learned a tremendous number of lessons from him. Moreover,

Figure 33.1. This picture was taken in the summer of 1986 as Zewail and I were contemplating the conversion of an old X-ray facility housing some diffractometers used by Prof. Linus Pauling into a femtosecond laser chemistry facility. The laboratory was ready by the end of November 1986 and we had results (the ICN experiment) by February 1987.

from reading his books and having attended his public lectures on the Middle East, I have learned about his concern for developing countries and science education in general. No question, Zewail was an outstanding scientist, mentor, and role model.

Author Biography

Marcos Dantus received B.A. and M.A. degrees in Chemistry from Brandeis University, and Ph.D. degree (1991), and Postdoc (1991–1993) at Caltech with Ahmed Zewail, where he helped to develop Femtochemistry and Ultrafast Electron Diffraction. These developments were recognized by the Nobel Prize in Chemistry in 1999. Dantus has been a professor at MSU since 1993 and is presently a University Distinguished Professor and the MSU Foundation Chair. He has pioneered the use of shaped ultrafast pulses to probe and control chemical reactions as well as for practical applications such as biomedical imaging, proteomics, and standoff detection of explosives. Dantus regularly collaborates with different branches of the US Department of Defense (DoD) and was invited to DARPA's Scientist Helping America, Arbitrary Waveform Generation, and Program for Ultrafast Laser Science workshops. Dantus has over 225 publications and is a Fellow of The National Academy of Inventors, The American Physical Society, and The Optical Society of America. Dantus founded Biophotonic Solutions Inc. in 2004 to develop and commercialize of an instrument capable of automated laser pulse compression. The MIIPS shapers are now enabling research around the world. In 2015, he founded MTBIsense LLC, a company with the mission of reducing concussion-related injuries in sports.

Chapter 34

BCH-codes are (In Fact) Good!

*Michael J. Collins**

At a Special Discourse at the Royal Institution on March 13, 1882, chaired by the then Prince of Wales (later King Edward VII), Eadweard Muybridge showed photographs that demonstrated exactly how a horse runs,[1] in contrast to the images that had been created by artists for centuries. The method that he employed to obtain them was as simple as it was original — as the horse ran through a sequence of threads, these in turn set off the shutters of a row of cameras lined up alongside. Just as ingenious was the way in which he reconstructed the motion by means of a zoopraxiscope, a primitive version of the motion projector.

Just over a century later, in March 1991, Ahmed gave a Friday Discourse at the Royal Institution in which he introduced his own work to capture the process of chemical reactions by describing Muybridge's concepts. I leave it to others to describe Ahmed's work and his great contributions but while, to the non-chemist that I am, chemical reactions, lasers, and the concept of a femtosecond (10^{-15} second) may disappear into a blur, crucially I could *visualize* abstractly what Ahmed had done when he

---

* michael.collins@maths.ox.ac.uk
[1] For a full account, see https://ia600206.us.archive.org/28/items/attitudesofanima00muyb/attitudesofanima00muyb.pdf.

reported on his trip at lunch to "The Round Table" in the Athenaeum at Caltech and described how he had started his lecture.

At the time, I was spending the academic year at Caltech as a Visiting Professor, and Rudolph A. Marcus had introduced me to that lunch group. I first met Rudolph when he was visiting Oxford in 1975–1976 as a Visiting Professor in Theoretical Chemistry following the death of Charles Coulson and as a Visiting Fellow of my own college, and we were neighbors in the house where I lived in a flat once occupied by William Beveridge, and he and his family lived on the ground floor. But it was at this table that I first met Ahmed. I was very much the young outsider — not an experimentalist, and a mathematician to boot — so it must have been a surprise to Ahmed when, since I was thinking about the structure of British universities and keen to learn what it was that was so special about Caltech and might be (and should have been) imparted to the British system, I asked if I could see his laboratory. Equally, and typically, he was delighted to have been asked, and took me there immediately after lunch. I still have his apparatus as a clear image in my mind and a little different from the glorified test tube with chemicals changing colour that I may have imagined before — as also the "Oxford Instruments" tag on his lasers. That said, what is it to a pure mathematician that Muybridge's exposure time was $2 \cdot 10^{11}$ femtoseconds!

Returning to memories of Ahmed, what abides most is that contrast between the simple — almost simplistic — ideas with which he could describe his work and the unerring complexity and novelty that he could then impart to it, the clarity of thought and purpose that lay behind all that he did, whether in science or in his role as a TIAA/CREF trustee, and his understanding of what makes a great academic institution work, namely the role that the academics should play in contributing to the policy decisions and shape of the institution in which they serve. These are epithets, too, that would apply to his description and vision of the political issues on the Middle East later as he became more involved with them. I have visited Caltech annually since 2004, and my visits have always been enriched by the opportunity to hear him on these and many other topics.

So, as a tribute to Ahmed, may I offer some mathematics, hopefully in a style to reciprocate what I learned from him and which I hope that he might have appreciated. If I learned anything from the nature of conversation at that Round Table, whether the topic be politics or science, or whether you had introduced it yourself or were expected to answer to an opening question from the formidable Francis Clauser who "presided" there for so many years, it was that you should be both clear and intelligible to all present, with their wide range of interests that might be represented on any given day. And what I shall cover here, as it happens, is probably easier to explain than my usual subject of group theory.

Scientific observation in Muybridge's time would have been faithfully recorded with pencil and paper — still the working medium for many mathematicians including myself! Now, data is transmitted digitally — namely by a sequence of 0's and 1's. Experimentalists worry about errors, and so here do mathematicians — but in a context different from making a mistake. You want to send digital information, but the medium introduces errors during transmission; how efficiently can you send "superfluous information" in order that you can detect and correct errors (as long as there are not too many) and so determine the original message (or data)? It is this question that lies behind the title of this paper and a small piece of unpublished work inspired that year by interaction with Bob McEliece, rewritten here with the hindsight of over 25 years (and the loss of the original information that had been confined to the 3.5 inch disks of yesteryear!).

Suppose that we divide the message into blocks (or words) of a fixed length n. Then the existence of a single error in a word can be detected by adding a single check sum digit to the end of that word,[2] but the error can be corrected only with knowledge of its position. Assume now that a word is received with exactly one error. How can we indicate its position? We will label the symbol positions $1, \ldots, n$ in binary notation. This requires the use of $\lceil \log_2(n+1) \rceil$ symbols, where $\lceil \ \rceil$ denotes the "smallest integer greater than or equal to," and we put $k = \lceil \log_2(n+1) \rceil$. We now append a suitable set of k check symbols to our original word and send a word of length $n + k$. In the received word, there are 2^k possibilities for the sequence of final k symbols, and we would like to use them to label the position of an error that has occurred within the first n symbols with, conventionally, $0 \ldots 0$

---

[2]The sum (modulo 2) of the n symbols. More generally, this determines whether the number of errors is odd or even (in that case, possibly zero).

indicating no error. Since $2^k \geq n + 1$, this is theoretically possible, although we have not described how we might choose the k check symbols; the concrete example in the paragraphs immediately below in fact shows that this can be achieved. So this does provide a target for the number of additional symbols required. By extending this argument, we might hope to be able to correct up to e errors in the original word with ek check symbols. Unfortunately, we would need to be able to correct errors in the additional places too; this can be done at relatively little extra "cost" by logarithmic repetition, and this was my approach to the problem at the time.

However, it is actually better to ask about "self-correcting" words. So let us start with a simple well-known example where again we will initially assume that at most one error occurs. We can think of words of length n as vectors in the vector space $\mathbb{F}_2^n$ where $\mathbb{F}_2 = \{0,1\}$ is the field of integers modulo 2. Suppose in particular that we take n to be of the form $n = 2^m - 1$ and write the integers $1, \ldots, n$ in binary notation as sequences of length m (where we fill in with 0's at the front if necessary). Now let H be the $n \times m$ matrix with these as the rows in order and use in our message only those words represented by vectors $\mathbf{c}$ for which $\mathbf{c}H = \mathbf{0}$.[3]

Now transmit such a word $\mathbf{c}$ and suppose that a word $\mathbf{w}$ is received. If there is no error, then $\mathbf{w} = \mathbf{c}$ and $\mathbf{w}H = \mathbf{c}H = \mathbf{0}$. If it is known (or assumed) that just a single error has occurred, then we can write $\mathbf{w} = \mathbf{c} + \mathbf{e}_i$ for some permissible "codeword" $\mathbf{c}$ and a vector $\mathbf{e}_i$ that has 1 in the ith place for some i and 0's elsewhere. All we need to do is to determine i, for then we can recover $\mathbf{c}$. If we compute $\mathbf{w}H$, we see that

$$\mathbf{w}H = (\mathbf{c} = \mathbf{e}_i)H = \mathbf{c}H + \mathbf{e}_i H = \mathbf{e}_i H = \mathbf{h}_i,$$

the ith row of the matrix H – and this represents the integer i in binary form! Thus we are able both to detect and to locate (and therefore correct) that single error. Since the set of codewords is given as the set of solutions of a system of m independent homogeneous linear equations, they form a space of dimension $n - m$.

---

[3] Here, as throughout, we have simplified our notation; the normal practice is to take the transposes of these matrices.

With a little reorganization, this example can be used to establish the claim of our earlier informal discussion, replacing n with $n - m$ and putting $k = m$. Also, we note that no nonzero codeword can have fewer than three nonzero coefficients, and that there are codewords with 1's in just places i, j, k precisely when $\mathbf{h}_i + \mathbf{h}_j + \mathbf{h}_k = \mathbf{0}$. In fact, this is an optimal situation as we will explain later.

So now let us look at the more general picture. A (binary) *block code* consists of a set of words each of length n in symbols 0 and 1. Such a block code C is a *linear code* if it is a subspace of the vector space $\mathbb{F}_2^n$, and those words in C are called *codewords*. The (*Hamming*) *distance* $d(x, y)$ between two words x and y in $\mathbb{F}_2^n$ is the number of positions in which they differ. The minimum distance d of a code C is just the minimum distance between any pair of distinct codewords. Since distance is translation invariant in a linear code, the minimum distance d is also the minimal *weight* of a nonzero codeword. If a codeword is transmitted and then received with e errors, then the nearest codeword to the word received is guaranteed to be that which had been sent whenever $e \leq \frac{1}{2}(d-1)$. Thus in the example above we have $d = 3$, and it is possible by this approach to correct one, but not necessarily[4] two, errors.

Suppose now that $\dim C = k$; we say that C is an $[n, k, d]$-*code*. Take a basis for C and write its vectors as the rows of a $k \times n$ matrix A; then by performing a sequence of operations that add one row to another, or interchange two rows, we can obtain its so-called reduced echelon form — the first nonzero entry of each row is the unique nonzero entry of its column, and to the right of the first nonzero entry of the preceding row — this turns out to be independent of the initial basis chosen. We may also take a fixed permutation of all the symbols in our code since this is distance invariant; thus without loss we may assume that these "leading 1's" occur in the first k places. The first k symbols of each codeword are now the *information digits* and the remaining $n - k$ symbols *check digits*. We seek a good lower bound for $n - k$ and then show that there are indeed codes that "get close."

---

[4] In fact, we can never correct more than one error in this example, but this is not obvious at this stage. We will explain why later.

In order to be able to correct up to e errors, we have to be able to use the set of check digits to detect all patterns of up to e errors or equivalently all words of weight at most e, representing the actual errors. But this number is simply the sum of binomial coefficients[5]

$$1 + \binom{n}{1} + \binom{n}{2} + \cdots + \binom{n}{e}$$

and we can rewrite this expression as a polynomial

$$\frac{1}{e!} n^e + \frac{(3-e)}{2 \cdot (e-1)!} n^{e-1} + \text{lower degree terms}.$$

In particular, provided that $e \geq 4$, the second term is negative and dominates the lower degree terms for n sufficiently large.

If $e = 1$, we have already seen that the bound $n - k \geq \lceil \log_2(n+1) \rceil$ is optimal, while for $e = 2$ we have $2^{n-k} \geq \frac{1}{2} n(n+1) + 1$ from which we deduce that $n - k \geq 2 \log_2 n - 1$. When $e = 3$, we obtain an inequality $2^{n-k} \geq \frac{1}{6} n^3 + \frac{5}{6} n + 1$ from which we conclude that $n - k \geq 3 \log_2 n - \log_2 6$.

Generically, we now fix $e \geq 4$ and look at the behavior of the corresponding bound for n large. We will comment on the significance of this assumption later. Then we have

$$\log_2 \left(\frac{1}{e!} n^e - \frac{(e-3)}{2 \cdot (e-1)!} n^{e-1} \right) = e \log_2 n + \log_2 \left(1 - \frac{e(e-3)}{2n} \right) - \log_2(e!)$$

$$> e \log_2 n - \frac{e(e-3)}{2n} - \log_2(e!).$$

This then leads to an asymptotic inequality

$$n - k > e \log_2 n - \log_2(e!) + O(n^{-1}),$$

---

[5] Since $\binom{n}{i} = {}_nC_i$.

where the constant involved in $O(n^{-1})$ depends on e alone.

Specifically, we have established the following.

Theorem A. Let $\mathcal{F}$ be a family of binary linear codes, all of which can correct any pattern of e errors for some fixed $e \geq 2$. Then, for any $\varepsilon > 0$, there exists an integer N such that if $\mathcal{C} \in \mathcal{F}$ is a code of dimension k and length $n \geq N$,

$$n - k \geq e \log_2 n - \log_2(e!) - \varepsilon.$$

For $e = 2$ or 3, the inequality holds for all n, while if $e = 1$ we have only that $n - k \geq \lceil \log_2(n+1) \rceil$.

As a lower bound, a precise analysis above would give a logarithmic form of what is known as the *sphere packing bound*, a purely combinatorial result about codes that are not necessarily linear. So in this sense, the result is not new, though to obtain it in the context only of linear codes is, but crucially what this approach has done is to show that the bound is still a sensible target for families of linear codes, and it is in this sense, that we will explain why the family of *BCH*-codes, which are generally regarded as "bad" from a general code-theoretic standpoint, should be viewed more favorably. And this explains the title of this note.

First, we note that our original crude estimate is not very far from the best possible as shown in this theorem. Next, we remark that there are two particular features in the asymptotic behavior of a family of linear codes as the length n increases that one might examine, the *transmission rate* k/n and the *error correction rate* e/n. The family is deemed to be good if both rates can be bounded away from zero; in this sense, *BCH*-codes are bad. But we are interested here only in k/n, or more precisely $n - k$, for any fixed e. With this in mind, we make the following definition.

Definition. Let e be a (fixed) positive integer. Let $\mathcal{F}$ be a family of binary linear codes, all of which are can correct any pattern of e errors. We say that $\mathcal{F}$ is e-**good** if there is a sequence $\{C_i\}$ of codes in $\mathcal{F}$ of increasing length for which, if C_i has length n_i and dimension k_i,

$$n_i - k_i \leq e \cdot \lceil \log_2(n_i + 1) \rceil.$$

Thus we are saying that an *e*-good family has a subfamily that gets very close to the lower bound given in Theorem A, and at the same time that the existence of *e*-good families shows that this bound is close to optimal.

We refer the reader to the classic text by MacWilliams and Sloane[6] for the details of what we now describe, and for the proofs of what we will claim.

First, when $e = 1$, we note that this definition includes the optimal situation described initially. But there is indeed then a family of codes that meets this criterion, namely the binary **Hamming codes** $\mathcal{H}_m$ of length $n = 2^m - 1$ ([MS], p. 23 ff.). These codes, further, are *perfect*; they meet the sphere-packing bound, that every word that is not a codeword is at distance exactly 1 from a unique codeword. And it was this family of codes that we described earlier: a sphere of radius 1 and center, a codeword contains $1 + n = 2^m$ vectors, distinct such spheres are disjoint because $d = 3$, and, since

$$2^{n-m} \cdot 2^m = 2^n = |\mathbb{F}_2^n|,$$

every vector that is a not a codeword is at distance 1 from exactly one codeword. Thus "closest vector" decoding can never correct more than one error.

Now to justify the title of this note! There is a family of codes known as the ***BCH*-codes** discovered independently by Bose and Ray-Chaudhuri (1960)[7] and by Hocquenghem (1959).[8] The binary *BCH*-codes have length of the form $n = 2^m - 1$. They are defined in a fashion analogous to the Hamming codes described in the previous paragraph, but by an $n \times m$ matrix H whose entries lie in the larger field of order 2^m with codewords (still with just 0's and 1's) defined by the relation $\mathbf{c}H = \mathbf{0}$. There are further notions of *narrow sense*, *primitive*, and *designed distance* δ which we will

---

[6] F.J. MacWilliams and N.J.A. Sloane, *The Theory of Error-Correcting Codes* (North-Holland, 1977) — specific references will be to theorems in Chapter 9, and indicated by [MS] Amsterdam.

[7] R.C. Bose and D.K. Ray-Chaudhuri, Further results on error correcting binary group codes, *Information and Control* **3** (1960), pp. 279–290.

[8] A. Hocquenghem, Codes correcteurs d'erreurs, *Chiffres* **2** (1959), pp. 147–156.

not explain here, except to say that they are involved in the algebraic nature of the choice of the matrix H, and that we will assume that our *BCH*-codes are now always primitive and narrow sense without comment. With that restriction, let $C(n,\delta)$ be a *BCH*-code of length n and designed distance δ. Then its key properties are the following (see [MS], Chapters 7 and 9).

$C(n,\delta)$ has minimum distance at least δ.

If $\delta = 2t + 1$, then $\dim C(n,\delta) \geq n - mt$.

If $\delta = 2t + 1 < 2^{m/2}$, then $\dim C(n,\delta) \geq n - mt$.

If $e \leq t$, then $C(n, 2t+1)$ is therefore e-error correcting. On the other hand, if we fix e and take $\delta = 2e + 1$, then for large enough n we have

$$n - \dim C(n, 2t+1) = me = e \cdot \log_2(n+1).$$

Thus we have established the following.

Theorem B. For any $e \geq 2$, the family of binary *BCH*-codes of designed distance $2e + 1$ is e-good.

BCH-codes are useful for two reasons. First, and crucially, their particular structure gives rise to the existence of good decoding algorithms based on the underlying algebra. More importantly, they do have a good practical use in situations where it is not unreasonable to assume that errors come in "small bursts" so that our assumptions are valid — imagine, by analogy, a gramophone record with a few scratches or other minor blemishes; as long as they do not all appear on the same successive grooves, they may be viewed in this way. For this reason, *BCH*-codes, or more often other codes based on them, have long been used for embedding data on compact disks.

We will make one further observation. The linearity of our codes was used to identify the concept of "information" and "check" symbols. Linearity also shows that each check digit is a fixed linear combination of the information symbols. After this, the methods were purely information theoretic, using the linearity to define a one-to-one map between certain error patterns and the space spanned by the check symbols alone. Thus similar arguments might be exploited if nonlinear functions were used to define the check digits.

What we have shown in Theorem B is that there is a "usable" family of linear codes that get close to one type of optimal bound. And "optimal" must surely mark Ahmed's life and work.

Author Biography

Michael Collins (b. 1945) is an Emeritus Professor of Mathematics at the University of Oxford and an Emeritus Fellow of University College. He obtained his degrees in Oxford, winning both Junior and Senior Mathematical Prizes, before spending a year at the University of Illinois at Chicago Circle. He was elected a Fellow and Praelector at University College in 1969, and was awarded a Lindemann Trust Fellowship in 1973, spending a year at the University of Chicago, followed by six months at the Institute for Advanced Study. Subsequently, he has held Visiting Professorships at Caltech, Ohio State University, and at the Universities of Chicago and Virginia, and in recent years has regularly been a Visiting Associate in Mathematics at Caltech. He is the author of *Representations and Characters of Finite Groups* (CUP, 1990) and has edited into book form *Finite Simple Groups II* (Academic Press, 1980) and *Modular Representation Theory of Finite Groups* (de Gruyter 2001). In 1990–1991, he held the office of Assessor in the University of Oxford, and he has been both Senior Tutor and Dean of University College.

Chapter 35

Ahmed Zewail: A Reminiscence

*Ahmed Okasha**

Gibran Khalil Gibran (The American Lebanese Poet) stated that friendship is a sweet responsibility and not an opportunity. I have enjoyed the friendship of the humane, altruistic colleague and scientist Ahmed Zewail for more than 20 years before his Noble Prize award.

The best investment in life is the everlasting loyalty in friendship. It is rare to find such dedication for the propagation of knowledge and science and solidifying the essence of values and morals. It is rare to find people with vision and clear specific targets pursuing its success with wisdom and consistency. We shall grieve Ahmed Zewail with our emotions, feeling gratitude for his dedication. We did walk in his funeral and burial but part of our brain refuses to bid farewell to him. This always happened for those who left an intellectual, scientific, and cultural impact in our lives.

The absence of those great men does not nullify their presence. They are alive while they are in their graves. The grief of separation from our hearts will be outlived in our minds, thoughts, and psyche. He was a great believer and a workaholic, sincere and conscientious in his work with emphasis on teamwork. His respect for his colleagues, juniors, and the old and young was greatly felt. He used to sacrifice in silence from his

---

*aokasha35@gmail.com

knowledge, health, and money. He believed in the ability of the youth; this was manifested in the selecting team in Zewail City for Science and Technology whether scientists, administrators, researchers, or lecturers.

On the banks of the Nile, the Rosetta branch, he lived an enjoyable childhood in the City of Desuq, which is the home of the famous mosque, Sidi Ibrahim. He was born February 26, 1946, in nearby Damanhur, the "City of Horus," only 60 km from Alexandria. In retrospect, it is remarkable that his childhood origins were flanked by two great places — Rosetta, the city where the famous Stone was discovered, and Alexandria, the home of ancient learning. The dawn of his memory begins with his days at Disuq's preparatory school. He was the only son in a family of three sisters and two loving parents.

The family's dream was to see him receive a higher degree abroad and to return to become a university professor — on the door to his study room, a sign was placed which read "Dr. Ahmed," even though he was still far from becoming a doctor. Culturally, his interests were focused — reading, music, some sports, and backgammon. The great singer Oum Kalthoum, the most famous and popular singer in the Arab World, had a major influence on his appreciation of music. After three decades, he still had the same feeling and passion for her music. In America, the only music he had been able to appreciate on this level was classical, and some jazz. Reading was his real joy.

As a boy, it was clear that his inclinations were toward the physical sciences. Mathematics, mechanics, and chemistry were among the fields that gave him a special satisfaction. Social sciences were not as attractive because, in those days, much emphasis was placed on memorization of subjects, names and the like, and for reasons unknown to him, his mind kept asking "how" and "why." This characteristic has persisted from the beginning of his life. In his teens, he recalled feeling a thrill when he solved a difficult problem in mechanics, for instance, considering all the tricky operational forces of a car going uphill or downhill. Even though chemistry required some memorization, he was intrigued by the "mathematics of chemistry." It provides laboratory phenomena which, as a boy, he wanted to reproduce and understand. In his bedroom, he constructed a small apparatus from his mother's oil burner (for making Arabic coffee) and a few glass tubes, in order to see how wood is transformed into a burning gas and a liquid substance. He always remembered this vividly, not only for the science but also for the danger of burning down their house! It is not clear why he developed this attraction to science at such an early stage.

After finishing high school, he applied to universities. He was accepted in to Alexandria University and to the Faculty of Science. Here, luck played a crucial role, which gave him the career he always loved most: science. He graduated with the highest honors — "Distinction with First Class Honor" — with above 90% in all areas of chemistry. With these scores, he was awarded, as a student, a stipend every month of approximately L.E.13, which was close to that of a university graduate who made L.E.17 at the time!

After graduating with the degree of Bachelor of Science, he was appointed to a University position as a demonstrator, to carry on research toward a Masters and then a Ph.D. degree, and to teach undergraduates at the University of Alexandria. This was a tenured position, guaranteeing a faculty appointment at the University. In teaching, he was successful to the point that, although not yet a professor, he gave "professorial lectures" to help students after the Professor had given his lecture. Through this experience, he discovered an affinity and enjoyment in explaining science and natural phenomena in the clearest and simplest way.

Dr. Zewail finished the requirements for a Masters in Science in about eight months. The tool was spectroscopy, and he was excited about developing an understanding of how and why the spectra of certain molecules change with solvents. Professors El Ezaby (a graduate of Utah) and Yehia El Tantawy (a graduate of Penn) encouraged him to go abroad to complete his Ph.D. work. All the odds were against his going to America. First, he did not have the connections abroad. Second, the 1967 war had just ended and American stocks in Egypt were at their lowest value, so study missions were only sent to the USSR or Eastern European countries. He had to obtain a scholarship directly from an American University. After corresponding with a dozen universities, the University of Pennsylvania and a few others offered him scholarships, providing the tuition and paying a monthly stipend (some $300). There were still further obstacles against travel to America. It took enormous energy to pass the regulatory and bureaucratic barriers.

His presence — as the Egyptian at Penn — was starting to be felt by the professors and students as his scores were high, and he also began a successful course of research. He was working almost "day and night," and doing several projects at the same time: The Stark effects of simple molecules; the Zeeman effect of solids like NO_2^- and benzene; the optical detection of magnetic resonance (ODMR); double resonance techniques, etc. Later, thinking about it, he could not imagine doing all of this again, but of course then he was "young and innocent." He applied for five positions, three in the US, one in Germany, and one in Holland, and all of them with world-renowned professors. He received five offers and decided on Berkeley. From his Ph.D. experience, he had gained work on the spectroscopy of pairs of molecules, called dimers, and how to measure their coherence with the new tools available at Berkeley.

He wrote two papers with Professor Charles Harris, one theoretical and the other experimental. They were published in *Phys. Rev.* These papers were followed by other work, and he extended the concept of coherence to multidimensional systems, publishing his first independently authored paper while at Berkeley. During this period, many of the top universities announced new positions, and Charles asked him to apply. He decided to send applications to nearly a dozen places and, at the end, after interviews and enjoyable visits, he was offered an Assistant Professorship at many, including Harvard, Caltech, Chicago, Rice, and Northwestern.

His interview at Caltech had gone well, despite the experience of an exhausting two days, visiting each half hour with a different faculty member in chemistry and chemical engineering. The visit was exciting, surprising, and memorable. The talks went well and he even received some undeserved praise for style. At one point, he was speaking about what is known as the FVH, picture of coherence, where F stands for Feynman, the famous Caltech physicist and Nobel Laureate. He went to the board to write the name and all of a sudden he was stuck on the spelling. Halfway through, he turned to the audience and said, "you know how to spell Feynman." A big laugh erupted, and the audience thought he was joking — he was not! After receiving several offers, the time had come to make up his mind, but he had not yet heard from Caltech. He called the Head of the Search Committee, now a colleague of his, and he was lukewarm, encouraging him to accept other offers. However, shortly after this, he was contacted by Caltech with a very attractive offer, asking him to visit with his family. We received the red carpet treatment, and that visit did cost Caltech! He never regretted the decision of accepting the Caltech offer.

His science family came from all over the world, and members were of varied backgrounds, cultures, and abilities. The diversity in this "small world" he worked in daily provided the most stimulating environment, with many challenges and much optimism. Over the years, his research group has had close to 150 graduate students, postdoctoral fellows, and visiting associates. Many of them are now in leading academic, industrial, and governmental positions. Working with such minds in a village of science had been the most rewarding experience — Caltech was the right place for him.

The journey from Egypt to America was been full of surprises. As a postgraduate student, he was unaware of the Nobel Prize in the way we now see its impact in the West. He used to gather around the TV or read in the newspaper about the recognition of famous Egyptian scientists and writers by the President, and these moments gave him and his friends a real thrill — maybe one day, we would be in this position ourselves for achievements in science or literature.

Some decades later, when President Mubarak bestowed on him the Order of Merit, first class, and the Grand Collar of the Nile ("Kiladate El Niel"), the highest State honor, it brought those emotional boyhood days back in his memory. He never expected that his portrait, next to the pyramids, would be on a postage stamp or that the school he went to as a boy and the road to Rosetta would be named after him. Certainly, as a youngster in love with science, he had no dreams about the honor of the Nobel Prize.

I was invited by Ahmed Zewail to deliver a short talk at the Gala Dinner in the Athenaeum in Caltech's Ahmed Zewail celebration on February 26, 2016. I started by stating: "It is both a privilege and an honor to address such a distinguished gathering of scientists and scholars. My remarks this evening will not address Ahmed's contributions to Science and to Caltech — you are a better judge! But rather to bring to your attention, a huge undertaking by Ahmed in his mother country, Egypt, now known as 'Zewail City of Science and Technology.'"

Ahmed Zewail, in his book *Age of Science*, describes himself as an Egyptian, Moslem, African, Asian, Middle Eastern, Mediterranean, and American. Being influenced by Pharaonic, Coptic, Islamic, and American cultures reinforced his telescopic perspective of the world. He was a man of integrity who belonged to his roots and had successfully assimilated into the American culture. He succeeded in having his scientific achievement in parallel to the service and welfare of societies, and the conference on "Science and Society" was a wonderful occasion for the celebration of both achievements.

All Egyptians identify with Dr. Zewail's success, especially after becoming the sole recipient of the Nobel Prize in science, in 1999. He became the hero of Egypt. I recall an incident that epitomizes the love of the people to him. After having a dinner in a Cairo restaurant on the Nile, we stopped to buy Ahmed the newspapers of the next morning. The vendor shouted: "Ahmed Zewail" and refused any payment from Ahmed or myself. "How do you know me? Do you know my work?" Ahmed asked the vendor. "I do not know anything; I only know that you are the pride of Egypt," said the vendor.

In Egypt, he was eager to launch a National Project for a renaissance of science, research, and development. President Mubarak assigned a new area of land on the outskirt of Cairo, and the foundation stone was laid on January 1, 2000, days after he received the Nobel Prize in Stockholm. After many obstacles and impediments, and a delay of 15 years, following the 2011 Egyptian revolution, a decree was issued for the establishment of "Egypt's National Project for Scientific Renaissance," and it was named "Zewail City of Science and Technology." One of the reasons for the delay in the implementation of Dr. Zewail's project was his high "Emotional Intelligence" and "charismatic personality." The masses considered him the savior of Egypt, and the mass media nominated him for the presidency of Egypt, both resulted in fears from authorities, which lead to halting of the progress of his project. A delay of 15 years for many would produce a sense of despair. However, the integrity of his personality, his resilience, dedication, persistence, and foresight were the driving force in his pursuit for the success of this endeavor. In Zewail City, he recruited the most eminent Egyptian scientists living abroad, attracted the best undergraduate students in all of Egypt, and hired the best administrative staff. Real scholars only care about achievements! Dr. Zewail's Project is structured to include a University, Scientific Institutes, a Think Tank, and the Hubs to house all scientific products for the welfare of the Egyptian society.

The purpose of the University is to build a new generation of leaders, scientists, and entrepreneurs capable of having a significant productive impact on the society. The university searches for talent and academic excellence from all over Egypt. After their first year, students choose between two major career options, Engineering or Science. The University offered new majors for the first time in Egypt and these include: Nanotechnology, Engineering, Renewable Energy and Environmental Engineering, and Space and Communication Technology. Fundamental sciences include Nanoscience, Biomedical sciences, Materials sciences, and Physics of the Earth and Universe. The aim of the Research Institutes at Zewail City is to carry out advanced research and development, especially in areas relevant to prevailing problems in the country. Of particular significance are areas relating to health, energy, and economic development. The goal is to create a sustainable development through industries, which are built on the output of scientific research carried out at the Research Institutes.

President El Sisy assigned 200-acres for a New Campus, and it is now being built with the help of the Engineering Authority of the Armed Forces. In the last e-mail sent to the Board from Dr Zewail, he stated: "We now have a new campus of 200 acres in October Gardens, and its inauguration is scheduled in the first half of 2016." He continues, "We now have more than 280 staff on board; we operate 7 research institutes with 120 academic staff; we have 530 top undergraduate students who had to achieve a national score of 96% and above; we produced 200 research papers presented in international conferences, and published 100 research papers presented in high-level international scientific journals." The Board of Trustees of Zewail City is composed of 30 world-class members, including 5 Nobel Laureates in various fields. We are honored that Dr. Tom Rosenbaum, the president of Caltech, has joined the Board beginning December of 2015.

Dr. Zewail was an exemplary person not only for the highest achievement and appreciation he obtained from numerous scientific organizations in the world but also for his services to humanity. Eric Fromm, in his book *To Have or To Be*, emphasizes that "To be" requires the need of belonging, the need to self-identify, the need of roots, the need to transcendence, and the need to live in a frame of futuristic goals. All these needs have been fulfilled by Dr. Zewail.

Dr. Zewail, you still have a pivotal impact on Science, Society, and Culture, not only in your home of scientific research, Caltech, but also by providing a global vision for a peaceful world —You are a Statesman!

Author Biography

Ahmed Okasha is Professor and Director of WHO Collaborating Center for Training and Research on the Mental Health at the Institute of Psychiatry, Ain Shams University, Cairo, Egypt. He is the President Egyptian Psychiatric Association, and an Honorary President of the Arab Federation of Psychiatrists. He served as the President of the World Psychiatric Association (2002–2005) and as the President of the Egyptian Society of Biological Psychiatry and WFSBP. He is a Member of WPA Council and has served on the editorial advisory boards of 23 international scientific journals. He is a Member of National Mental Health Council, Member of Supreme Council of Culture, and a Member of Supreme Council of Health. Prof. Okasha has published more than 285 original articles in national and international journals. He is the editor and contributor of 47 national and international books in the field of psychiatry and psychology in both the Arabic and English languages. He is a member and editor of many international psychiatric journals, an International Distinguished Fellow of the American College of Psychiatrists and WPA, and a Fellow of Royal College of Physicians (Edinburgh) and Royal College of Psychiatrists (London). He is also the recipient of Presidential Commendation from the American Psychiatric Association and honorary doctorates and fellowships from many Universities and Psychiatric Associations. He was awarded, by the Egyptian Academy of Science, the highest honors in Egypt: i.e., State Merit Prize in Creativity in Medicine (2000), State Merit Award in Medical Sciences (2007), Nile (Moubarak) Merit Award in Medical Sciences (2010), and Medal of Science and Arts of the First Degree (2013). He is also a member of the Egyptian Presidential Council for Distinguished Scientists (Mental Health & Community Compatibility).

Chapter 36

Goodbye My Love

Dema Faham*

As I kissed you goodbye one last time and you were letting go of my hand, our lives together flashed in front of my eyes.

I remembered how we met. I was entering the reception room at the Khozama hotel in Riyadh, Saudi Arabia, when I heard that infectious laugh of yours. You instantly grabbed all my attention. You told me that when you saw me, the whole room melted away and you only had eyes for me. I know it sounds unreal, like a fairy tale, but it really was love at first sight. I remembered how you sent me Oum Kalthoum love songs, expressing your feelings through her words. I thought about how we hit it off from the first phone call, talking for six hours straight as the time passed us by, unnoticed.

You courted me using every trick in the book. Maybe I should have let you know you did not have to try so hard — you had me from that first laugh. After six months, you asked me to marry you. When I told you it was too fast, you replied that six months was an eternity for someone who works in Femtoseconds.

We got married and had a lovely reception at the Athenaeum, where I was introduced to the Caltech community. It was also there that I fell in

---

* d.faham@sbcglobal.net

love with your two wonderful daughters Maha and Amani, who quickly became my best friends. You were very appreciative I made our house a home for them. You grew closer to them. We became a family.

Shortly after we married, I realized how dedicated you were to your work. One night after dinner, you called the lab and asked them to call you as soon as they got any results. Sure enough, at 2 a.m., our phone rang, and you absolutely could not wait until the morning. You asked Martin Gruebele — Professor at U of I — to bring the results to our home right away. You sat, waiting anxiously for him to arrive. Once he arrived, both you and Martin were excited about the results and you enthusiastically shared it with me. I looked at it, and saw a squiggly line on the paper. Why couldn't this fuss wait till morning! But there was no waiting with you. You were like a kid in a candy store; forever the enthusiastic young vibrant spirit.

You were always very competitive. It was clear to me from the beginning, based on all the stories you told me, about how you never accepted anything less than being at the top of your class. I had a taste of that one time when I was sick and feeling down. In an effort to cheer me up, you suggested we play a game of backgammon. I declined saying you would surely win, and I wasn't in any mood to lose that day. But you insisted it would be fun, and so I agreed on one condition: You would lose to me. You tried to wiggle out of it, arguing I wouldn't like winning knowing I hadn't earned it, but in the end you conceded grudgingly. I shouldn't be surprised you ended up winning. To your credit, you were very apologetic when I was obviously very upset, saying "I tried, I really did, but I just couldn't." I knew then that my husband couldn't stand being anything but on top in whatever he does. You were a perfectionist and wouldn't settle for anything but the best.

Three years passed before we had our first boy, Nabeel. His birth prompted you to spend more time away from the lab so you could spend more time with him. One year later, we completed our family with the arrival of Hani. You frequently insisted we put him in our bed as you loved cuddling with him (he was "squeezable!").

As your commitments took you around the world, and you were fully immersed with your work at Caltech, it became clear that you had little time to help with the day-to-day management of the kids. I decided to halt my career to raise our children, and you were so appreciative that I freed you to pursue your passion. You only needed to worry about pushing the frontiers of science, knowing that I was handling all the other aspects of running our household.

You made good use of all those robes and regalia that you collected from around the world. When you received various honorary degrees, you used them as costumes when you took the kids around the neighborhood trick or treating on Halloween, as they walked, proudly holding hands with "The Master."

You were happiest sitting in the backyard, drinking a cup of tea, and smoking a cigar — despite my protests — as we discussed world affairs, education, your beloved Egypt, and the politics of science. It was the best time to approach you if one of us had a question or was struggling with resolving a problem. Not only did you give us all the time we needed, but you listened attentively. You had that gift of summarizing what we presented you in an eloquent way, and then reorganized our scattered thoughts in such an elegant fashion that the answer presented itself, simple and clear. You had an amazing ability to cut through the complexity and zoom to the core simple truth. For that, everyone sought your advice. But that never stopped us from debating you fiercely, whatever the topic was.

We spent endless hours talking about your dream to build a science base in Egypt. You were always concerned about the fate of your Motherland, and the Arab world as a whole. Nevertheless, you remained forever hopeful that with the right guidance, we would all move toward a brighter future.

You were the romantic one. You loved Valentine's Day. Every year, I got roses with a nice gift and a beautiful card. When I told you it was a silly holiday, and I refused to be dictated to about when and how I need to express my love, you would say "I want you always to know that your love sustains me, your words guide me, and although the world may admire me, you are my world."

You looked forward to our family dinner every evening. As Nabeel got increasingly interested in the world of politics, and started competing in high school speech and debate, I remember how proud you were as he collected more trophies than we had shelf space. More importantly, he got to educate all of us about what transpired in D.C., and opened our eyes on many issues that were plaguing the world. You were most proud that you started to learn from him, and even seek his advice on varieties of topics. Hani used to provide us with humor, and lighten the mood whenever the discussion got heated. You always said that he was the clever and smart one, maybe even too smart for his own good. You both shared a love of music, and you used to enjoy when he played the Arabic tunes you liked. He also introduced you to Santana and Jimi Hendrix, but he failed to get you to listen to rap music despite his many attempts.

The whole family fondly remembers our Thanksgiving dinners. As I was putting all the food on the table, and getting ready to carve the turkey, everyone was engaged in a number of conversations. You always managed to turn the several conversations going on simultaneously into a "coherent" one. I remember you asking my five-year-old nephew, "What are your thoughts about light?" and quizzing my 10-year-old niece, "Why time doesn't go in the opposite direction?" You always knew how to ask the question, and to include everyone. Nothing was off limits in the discussion, and you made everyone feel that their opinion was valuable regardless of how trivial it was. In the end, we would talk, laugh, and eat so much that we would end up on the couch resting and watching TV, just trying to recover!

I remember how proud you were when any of the children got any recognition or award. On the other hand, if they brought home a less-than-perfect grade, you always rigorously questioned their efforts, and ask if they really put in a 100% effort. If you ever found them happy with a mediocre performance, you made sure they knew you thought they were being lazy. When you were asked if you would try to guide the kids in choosing a career in science, you would jokingly reply, "I am going to give them total freedom to choose what they fancy as long as it is either physics, chemistry, or astronomy."

We always knew, though, you were so proud of them. Who could forget how you were beaming when you handed Maha her degree when she graduated from Caltech, or how you were teasing Amani after she got accepted to Chicago Medical School that you need to inform her patients that she could not even cut an avocado properly. And when they chose a path that was not what you envisioned, like when Hani decided to leave UC Berkeley to pursue a career in music, or when you questioned

Nabeel's decision as he left a lush job at PIMCO after he graduated from Georgetown to join a start-up company, you always encouraged them and supported them in whatever endeavor they pursued.

Your faith and belief in God always sustained you. It provided you a sense of purpose and gave you serenity. I remember how we talked about the timeless debate among scientists about the existence of God, and how you were always in the camp that said the more we understand of the universe, the more it strengthen our belief in the existence of an absolute being. Of a divine nature that set the balance of the laws of physics that govern the world, and the intricacy of the universe.

Although the world is getting more connected via the internet and all the social media that made it possible, there are still many scholars who believe there will forever be a conflict between east and west, and a clash of civilizations is inevitable. To the contrary, you always believed that there was only a lack of opportunities — that if it were to be provided for the have-nots, then conflict would be resolved and the world would all benefit, and become richer. You said, "I am a Muslim. I am Egyptian. I listen to Oum Kalthoum. I love pizza. I live in Pasadena. I am a professor at Caltech, and I work on the most advanced scientific endeavor. No conflict!"

You were so committed to your group members that you wouldn't rest until you secured them a job, and a chance to move forward in their careers. You were always there for them — your door was always open if they needed advice, or sometimes a scolding. Even after they moved on with their lives, you were always there for them when they reached out to you. Over your long and illustrious career, you had more than 400 students and scientists pass through your lab, and you were forever proud of what you achieved with them. You considered their success your biggest accomplishments.

People who know you knew not only you loved Caltech, but you embodied what Caltech was about. You walked the campus proudly. It was the place where the sky was the limit for your imagination, where you joined your colleagues at the famous round table at the Athenaeum for lunch and had a stimulating discussion. It was a classy place where you could phone the VP or the Provost to resolve an issue, no memo needed. It was a welcoming place where they painted the name of the assistant professor on the designated parking spot the day he arrived. It was a "civilized" place where collaboration and helping was the norm. It was where everyone felt like family. And you worked so hard to maintain it that way whenever you felt something was slipping — from the changes to the transportation department, to the big science project, you always looked at Caltech as the ideal place.

You wanted to repair the relationship that Linus Pauling had with the institute, as he was not only an icon for you but you felt he was one of the pillars of Caltech. You were so pleased when Caltech not only opened their arms welcoming him back, celebrating his birthday with a historic conference, but also when you led the inauguration of the Linus Pauling Lectureship at Caltech. You were so proud when you were named the first Linus Pauling Chair.

I remember when you got the Max Planck Institute offer and you were tempted, as funding big ideas was getting harder, and you felt that you were held back. It was an opportunity to work on your big idea, no need to worry about resources or writing grants. You would have gotten a big team and could guide the research in any direction you wished. But your heart was at Caltech and neither one of us wanted to leave Pasadena. You agonized over the decision, and luckily Caltech was able to come through to help in supplementing some funds to rehabilitate your labs. In the end, you were able to stay at your beloved Caltech.

I remember how many invitations you received from different companies to sit on their boards, and although the financial compensation was handsome, you were never tempted by it. You felt it was a distraction from your true passion — science — and an invasion on your time, which was so valuable that you didn't want to waste those precious hours. But when the White House asked you to join the Presidential Council on Science and Technology (PCAST), you felt it was your duty to help guide the direction of education in the country, and we all encouraged you to jump at the opportunity, knowing it was a chance to put everything we had envisioned for a brighter future into action. And when President Obama chose you to be a science envoy to the Middle East, we were so happy because you felt that would open so many doors, and push you farther along on your dream to lift that part of the world through building bridges of knowledge.

It was the rare occasion that we took a vacation. You were always in a race with the time, trying to accomplish all you could. You were always so ambitious — reaching higher, moving faster. There were so many things you wanted to do, and it was as if you knew there would never be enough time to get it all done.

When you were diagnosed with Multiple Myeloma, it wasn't beating the disease that concerned you, but whether it would slow you down from accomplishing what you wanted to get done in Egypt, and getting your latest project up and running. You were in a race against time.

From the hospital bed, we called people to solve an arising problem. From the infusion chair, we edited documents they sent you. Despite the fatigue and the pain you were experiencing, you kept pushing yourself to ensure a smooth transition, and instead of resting and taking time off, you were doubling your workload. You asked me to take care of all the medical decisions and your entire treatment schedule so you could concentrate on the needs of Zewail City. You neglected doctors' advice to stay put and avoid traveling to Egypt during treatment. You were more determined to see it through, and despite the fact that every time you returned from Egypt, you had to be hospitalized due to a virus or some other illness you picked up there, you were adamant and kept pushing through. I helped you knowing that it was the fuel that kept you going. And the Egyptian people saw that. They love you for it, and will stay loyal to your memory, and the dream you dedicated your life to.

I was looking at you there in the hospital bed. The doctors came in the room, saying they did all they could. Saying they were sorry. I told them, "Don't be. He lived every minute, and femtosecond, of his life exactly how he wanted. He had no regrets. He did more than ten people combined could do in one lifetime. He lived his life to the fullest, and went out smiling."

Goodbye, my love. You will live in my heart forever. Your laugh will always ring in my ear. Your spirit will lift me when I am down, guide me when I am lost, and give me strength when I am weak. Be assured your children will always look up to you as their hero, and they will always thrive to sustain your pride in them. Every member of the family misses you, and you will always occupy that special place in our gathering. Your friends will cherish the memories they shared with you, your students will always be inspired by you.

You will be remembered for your determination, your optimism, your science, and your contributions to humanity. I love you, Ahmed.

Dema and Ahmed were married 27 years. Together, they raised two children — Nabeel and Hani.

Additional List of Obituaries

Chergui, Majed

"Ahmed H. Zewail," *Phys. Today*, web version. Available at: http://physicstoday.scitation.org/do/10.1063/PT.5.6244/full/.

"Ahmed H. Zewail," *Phys. Today* **69**, 69–70 (2016).

"In Memoriam: Ahmed H. Zewail (1946–2016)," *Struct. Dynam.* **3**, 040401 (2016).

Dervan, Peter

"Ahmed H. Zewail (1946–2016)," *Science* **353**, 1103 (2016).

El Sayed, Mustafa

"Ahmed H. Zewail: A Force for Egyptian Science," *Proceedings of the National Academy of Sciences*, **114**, 1743–1744 (2017).

Salmawy, Mohamed

See: http://weekly.ahram.org.eg/News/17039/21/Zewail-and-Mahfouz.aspx.

Thomas, John M.

"Ahmed H. Zewail (1946–2016)," *Angew. Chemie. Int. Edn.* **55**, 11335–11336 (2016).

"Zewail's Lasting Legacy," *Al-Ahram Weekly*. Available at: http://weekly.ahram.org.eg/News/17065.aspx.

Warren, Warren S.

"Ahmed H. Zewail (1946–2016)," *Nature* **537**, 168 (2016).

www.ingramcontent.com/pod-product-compliance
Lightning Source LLC
Chambersburg PA
CBHW071018240526
45469CB00006BD/1974